T0335592

Applied Mathematical Sciences

Volume 86

Applied Mathematical Sciences

(continued following index)

Dimitrios S. Dendrinos Michael Sonis

Chaos and Socio-Spatial Dynamics

With 49 Illustrations

Springer-Verlag
New York Berlin Heidelberg
London Paris Tokyo Hong Kong

Dimitrios S. Dendrinos
Urban and Transportation
Dynamics Laboratory
2011 Learned Hall
University of Kansas
Lawrence, KS 66045-2962
USA

Michael Sonis
Department of Geography
Bar-Ilan University
Ramat-Gan 52900
Israel

Editors

Mathematical Subject Classifications: 58F, 92A, 34G, 35F, 35G, 35K, 58E, 76F, 90A, 34C, 34D, 34K, 35B, 35E, 35J, 35R, 47H, 58C, 76E, 90B

Library of Congress Cataloging-in-Publication Data
Dendrinos, Dimitrios S.
Chaos and socio-spatial dynamics/Dimitrios S. Dendrinos, Michael Sonis.
p. cm.
Includes bibliographical references (p.).
ISBN 0-387-97283-8
1. Population geography—Mathematical models. I. Sonis, Michael. II. Title.
HB 1951.D46 1990
304.6—dc20 90-9672

Typeset by Asco Trade Typesetting Ltd., Hong Kong.
Printed and bound by R.R. Donnelley & Sons, Harrisonburg, Virginia.
Printed in the United States of America.

9 8 7 6 5 4 3 2 1

ISBN 0-387-97283-8 Springer-Verlag New York Berlin Heidelberg
ISBN 3-540-97283-8 Springer-Verlag Berlin Heidelberg New York

To the memory of Henry Mullally

Preface

The main objective of this book is to present some key qualitative features of a universal discrete relative dynamics map (iterative process). It is argued that the form of this map can accommodate a wide range of dynamics found in social systems distributed in (discrete) space. Particular emphasis is placed on population dynamics.

A gamut of models are included in the book, ranging from the one-stock, two-location case to multiple-stock, multiple-location examples. It is shown that the qualitative properties of these dynamics vary significantly as one spans the spectrum of the models studied. Many novel phenomena reproduced by the map are still awaiting to be uncovered. Among the innovative qualitative features presented here, those of interest are the two-dimensional discrete dynamics bifurcation, which is equivalent to the Hopf bifurcation found in continuous, two-dimensional dynamics; the onset of local and partial turbulence involving the initial phase of period-doubling cycles only; the existence of a variety of strange attractors, and what we also call "strange containers" of chaotic motion; etc.

Examples are also supplied of how the universal map can be used in socioeconomic–geographical systems. An advantage is enjoyed by this universal map vis-à-vis other nonlinear models. Through appropriate logarithmic transformations, linear regression techniques can be employed to test specific hypotheses regarding spatial population (or other social stock) dynamics.

The book is divided into four parts. Part I contains a brief introduction to an abstract theory of socio-spatial evolution set in a discrete time–space mode and expressed as nonlinear dynamics of stocks. It sets the stage for presenting our universal discrete relative map.

In Part II the simplest version of our mapping is discussed in the context of the one-stock, two-location model with a log-linear and a log-log-linear specification. Although the analysis draws from classical location theory, it paves the path for a comprehensive (i.e., composite) view of socio-spatial forces and their effects. Analytical and computer simulation results are presented, and some new qualitative features in discrete dynamics are outlined. This part contains the mathematically innovative elements of the work, including the discrete dynamics Hopf bifurcation equivalent of continuous dynamics; and the appropriate specifications of the Feigenbaum sequences, which are attributed to slopes rather than parameter values.

Parts III and IV contain the one-stock, multiple-location and the multiple-stock, multiple-location dynamics, respectively. The generalized discrete-time logistic growth for the one- and two-stock, multiple-location model is also introduced. In Part III an example is provided of how this universal map can be used in socio-spatial dynamics. Empirical evidence is presented, describing U.S. regional relative population distributions according to the universal map, covering approximately two centuries of observations and forecasts. Finally, in the Epilogue, a few central issues–events associated with this dynamic process are explored, as to how they may be of direct use to the social sciences. A number of examples are supplied where this map extends the existing models of socioeconomic–geographic systems.

This book is intended for use as a resource by researchers in the field of applied mathematics and all fields of social, biological, and natural sciences. The rigor of the mathematical exposition remains at the social scientists' level, as the qualitative dynamical (and innovative) features are simply exposed awaiting the mathematician's rigorous proof. The book can be used for advanced graduate courses in applied mathematics, with a focus on any field of the social sciences and humanities emphasizing spatial analysis.

Acknowledgments

Work on this book started in early 1984. Since then, numerous elements of it have been presented at various scientific conferences and seminars, at a number of universities in North America, Europe, Australia, and the Far East. We wish to acknowledge the contribution of those scientists—it is not possible to mention them all here—who have provided many useful remarks on this work during those presentations.

Henry Mullally and Takatoshi Tabuchi have read versions of the entire manuscript and provided many helpful comments and suggestions. Of course, any remaining errors or omissions are solely our responsibility.

Various papers, partly covering sections of this book, have appeared in the *Journal of Applied Mathematics and Computation*; *Mathematical Modelling: An International Journal*; and two books, *International Perspectives on Regional Decentralization*, Nomos, Baden-Baden, 1987, and *Economic Evolution and Structural Adjustment*, Springer-Verlag New York, Inc., Lecture Notes in Economics and Mathematical Systems, Vol. 293, 1987. To their editors, J. Casti, X.J.F. Avula, I. Masser, P. Freidrich, D. Batten, and B. Johansson, we extend our appreciation for their support.

Figures 20, 22, 24, 41, 42, and 43 are reproduced here with permission from Elsevier Science Publishers.

We express our thanks to the Transportation Center of the University of Kansas for making their limited resources available to us in the production of the manuscript. To our editor with Springer-Verlag New York, Inc., Rüdiger Gebauer, his assistant Susan Gordon, and their staff we extend our gratitude for supporting the publication of this volume and for their high-quality work in producing this book.

All numerical simulations, and a great deal of the theoretical part of this work, were developed under contract from the National Science Foundation (SES-82-16620 to D.S. Dendrinos), which is gratefully acknowledged, although the opinions expressed here do not necessarily reflect those of the NSF. However, all parts of this work were uncovered and developed by both authors in close and continuous cooperation since 1983. The name sequence on the cover is simply alphabetical.

Lawrence, Kansas — DIMITRIOS S. DENDRINOS
Ramat-Gan, Israel — MICHAEL SONIS

Contents

List of Figures

List of Tables

List of Tables

Prologue

This book presents a powerful, and rich in dynamical behavior, model which describes the evolution and spatial distribution of populations on the basis of the temporal and locational advantages enjoyed. A discrete time–space universal map of relative dynamics is used in unfolding a large catalogue of evolutionary events, some not previously discussed in either the mathematical or social science literature. Calm and chaotic events in a geographical–historical context are postulated.

Systems are the way they are because they become what they are. This is a well-known dictum in physics. The dictum is also informative in theorizing about socioeconomic systems. Dynamical analysis is at the core of socioeconomic behavior. In this book, we attempt to incorporate some recently developed maps of discrete, relative, nonlinear interactive dynamics into the analysis of the socioeconomic systems evolution in space–time.

We explore in detail the analytic properties of a universal map of relative space–time dynamics. This map is broadly applicable to socioeconomic–geographic systems, as is the simple logistic dynamic model widely applicable to biological and social systems.

Socioeconomic systems are characterized by a menu of dynamical phenomena: frequently, some systems are regularly periodic, like the daily congestion events during peak hour travel on urban highways. At times, nonregular but periodic oscillations describe business or economic events: the business cycle, annual budget cycles, and seasonal variations in various indicators of economic performance are examples of such oscillatory behavior. Nomadic population movements are also examples of cycles extending in space–times.

Longer term nonregular, periodic movements have been observed in the growth and decline of industries, cities, and national economies. However, by far the most frequently observed phenomena in socioeconomic systems are turbulence and chaos. A very striking example of such events is the behavior of stock (or commodity) prices in stock (or commodity) exchanges. Turbulent and chaotic behavior is found in a broad array of socioeconomic–geographic events, ranging from psychological (individual behavior) to political and demographic (mass behavior) variables. Thus, the study of dynamical systems capable of reproducing such violent behavior is of interest to social scientists.

Simple dynamic models have been shown to be a helpful means toward providing new understanding and insight into socioeconomic spatial dynamics. They reveal possible new events to look for, as well as uncovering previously unaccounted for phenomena in the dynamic (qualitative and, at times, quantitative) behavior of social systems. An analyst can interpret, from new and interesting angles, past and present socio-spatial incidents.

Recent developments in difference and differential maps have triggered profound developments in physics, chemistry, and biology. They seemingly hold a similar future for the social sciences. Nonlinear dynamics and their qualitative features may provide the means (elusive so far) of unifying the natural, biological, and social sciences.

From an analytical standpoint, the fixed-point process we present here is a universal map of discrete *relative* stock dynamics. The model has three main advantages when compared with other dynamic models. First, it is a general model which could be employed in a variety of contexts, exactly like the simple logistical prototype $x(t + 1) = Ax(t)[1 - x(t)]$, to simulate dynamic stock behavior. Second, due to its log-linear specifications, it renders itself to standard linear least squares regression techniques; this feature enables the analyst to easily test specific dynamical hypotheses in socioeconomic spatial behavior. Third, the mapping can be disaggregated among stocks and locations; in this book we explore the behavior of the universal map for the two- and multiple-location, one-stock problem; and the multiple-location, multiple-stock cases. A lesson demonstrated by this analytical exercise, is that the level of stock or spatial disaggregation considered, significantly affects the qualitative dynamic properties obtained by the universal mapping.

A number of novel (to the mathematical literature) and interesting (to the socioeconomic sciences) dynamic features are exposed. They include, among others, the following findings: three fundamental, discrete-time, discrete-space, dynamic bifurcations; one of them is the two-dimensional discrete analogue of the Hopf bifurcation found in continuous dynamics, which transforms a fixed point (attractor) to a stable two-period cycle. Other dynamic events include stable and unstable competitive exclusion; deterministic turbulence not obeying, in parameter sequences, the Feigenbaum universal constants, or the Li–Yorke condition that necessarily "a three period cycle precedes chaos," "strange attractors" and "strange containers;" "role reversal;" sequences of extreme positions with violently switching states; stocks vacillating between a conversion toward competitive exclusion, to states of complete dominance; the onset of "partial" and "local" period-doubling turbulence, as well as "complete" and "global" turbulence in multidimensional systems, etc.

The universal map presented here thus replicates calm, periodic, quasi-periodic, and violent or chaotic events in the socioeconomic systems' spatio-temporal evolution. In work following the writing of this book, and through computer simulations, it became evident that the phenomenon of quasi-periodicity (see Schaffer, Truty, and Fulmer, 1988) is widespread, particularly in the three-location (and higher-dimensional) cases of the universal map.

This work will be reported in forthcoming publications. We set out to test empirically the universal map for the case of regional population dynamics. Results from one experiment, involving the observed relative population distribution dynamics of the United States' nine divisions from 1850 to 1980, with projections to the year 2050, are briefly reported to demonstrate the map's extent of applicability.

Undoubtedly, there are many facets to this map, analytical and substantive, which are not addressed here. Future work, by us and other interested researchers, could possibly reveal yet richer elements embedded in the universal map presented here.

Part I
Socio-Spatial Dynamics

A. Introduction

We open the presentation of a universal mapping in discrete-time, discrete-space relative stock dynamics, the main subject of this book, by supplying a brief survey of the recent developments in the field of mathematics, the natural sciences, biology, mathematical ecology, and the spatial socioeconomic sciences. These developments provide the background for presenting the geographical applications of our universal map.

1. *A Brief Look at the Literature*

Influential in the development of socio-spatial dynamic theory have been the recent, as well as the relatively old, researches in mathematics in connection with the theory of simultaneous differential equations, bifurcation, and catastrophe theory. Among them, particular impact has been made by the work of Volterra and Lotka (found in Scudo and Ziegler (1978)), especially the work by Volterra (1927) on mathematical ecology; the work on the Hopf bifurcation (see, e.g., Marsden and McCraken, 1976); Lorenz' (1963) work on bifurcation theory in simultaneous differential equations; the Thom (1975), the Zeeman (1977), and the Poston and Stewart (1978) work on catastrophe theory; the work by Nicolis and Prigogine (1977) on self-organization in physical chemistry; Haken (1977) on synergetics; May (1974), (1976) on mathematical ecology, complexity versus stability, and the simple logistic prototype model; Feigenbaum (1978) on certain universal constants found in period-doubling, turbulent and chaotic behavior of dynamical systems, etc.

For an early survey of the bifurcation literature, which proved to have played an important role in the recent developments in socio-spatial dynamics, the reader is directed to May and Oster (1976). Earlier work of importance also includes Scott, Chu, and McLaughin (1973), Smale (1973), Li and Yorke (1975), among others.

Developments in the social sciences, economics, sociology, geography, etc., have been profoundly affected by the recent discoveries in mathematics and the natural sciences. Notions of continuous or abrupt growth and decline in socioeconomic spatially distributed stocks; extinction and morphogenesis; phase transitions; disequilibrium dynamics; cycles, periodic movement, turbulence, and chaos, are a few of the new concepts introduced into the theories of socio-spatial evolution.

These developments have been fueled by, among other factors, significant and interesting changes currently underway in human societies. Of central importance in these events are: the very recent and rapid rise in the size of the world's largest urban agglomerations, particularly those of the Third World; the sudden growth and decline of industries in the developed nations; sharp rises and falls in the prices of natural resources in international markets, and the effects they entail for nations dependent upon these commodities; steep rises in, followed by collapse of the world's stock markets, etc.

Over the past decade, a very large number of reports and papers (by many authors, not listed here) have appeared in the socioeconomic sciences in general (sociology, economics, and political science), and in the spatial sciences specifically (urban geography and regional science). Three books provide the link between socioeconomic, spatiotemporal dynamics and the earlier cited works. They are (in chronological order of appearance). A.G. Wilson's (1981) *Catastrophe Theory and Bifurcation: Applications to Urban and Regional Systems*, W. Weidlich and G. Haag's (1983) *Quantitative Sociology: The Dynamics of Interacting Populations*, and D.S. Dendrinos' (with H. Mullally) (1985) *Urban Evolution: Studies in the Mathematical Ecology of Cities*. All three contain extensive bibliographical references to earlier books, papers, and reports by a variety of authors in both the original fields (mathematics, the natural sciences, and biology) and in geography, economics, sociology, etc.

In each case the links to the background work differ. A.G. Wilson's work links, through the papers by Poston and Wilson (1977) and Harris and Wilson (1978), catastrophe theory and retail related locational interaction dynamics. They do not, however, employ the analytical foundations of catastrophe theory, namely its gradient structure, but largely rely on its qualitative features regarding discontinuities. Weidlich and Haag's work links stochastic master equations and the Fokker–Planck equation (developed in elementary particle and laser physics), with the theory of population migration. The A.G. Wilson, and Weidlich and Haag models contain spatiotemporal interactions.

The work by Dendrinos and Mullally links socio-spatial population dynamics to Darwinian evolution, through the application of the Volterra–Lotka formalism and the work by May found in the field of mathematical ecology. It is within this last vein of work that this book is undertaken. It further attempts to link work on turbulence and bifurcation theory in mathematics, physics, and mathematical ecology with socio-spatial (human population) dynamics.

Although the specific definition of the term "evolution" in social systems is still in its formative stage, the notion seems to imply at least some change of state in their behavior over time. Such structural change may involve smooth or abrupt transition in the main dynamic qualitative properties of the system at hand.

At the outset it is noted that socioeconomic phase transitions and bifurcations result from dynamic feedback processes among stocks distributed in space–time. These transitions affect the size, location, and interdependencies (forces, linkages) among the socioeconomic stocks. Whether, in general, any socioeconomic abstract quantities are preserved under these transitions in the spatiotemporal paths of the socioeconomic stocks is still an open question. In some specific cases, involving relative stock size spatial dynamics, there seem to be quantities which are preserved. An example includes the presence of a governing potential in relative spatial dynamics (see Dendrinos and Sonis, 1986), equivalent to that of Volterra's conservative ecologies.

Socioeconomic spatiotemporal dynamics are built on the assumption that, in spite of the multifaceted, very complex, individual, game-theoretic behavior, the aggregate performance of social stocks can be captured by simple spatiotemporal dynamic models. But the very simple dynamic models describing the aggregate performance of socioeconomic systems turn out to be dynamically complicated—and surprising—in the variety of events they reveal.

Socioeconomic system dynamics recognize also that social events have explicit geographical features. Their structure and functon are defined in and by space, as well as in and by time. In the model presented here, it is explicitly stated that space and time are of importance in the evolution of socioeconomic systems because both are *heterogeneous.* No two locations or time periods are alike. Thus, at least the spatial aspect of this work differs from the conventional geographical and regional science work based on von Thunen (1826), Christaller (1933), Lösch (1937), Hotelling (1929), Isard (1956), Alonso (1964), and others. The latter work recognizes space only through an impedance measure (transportation cost), whereas space is considered homogeneous (isotropic) otherwise.

So far, the focus of these new endeavors has been on population (demographic) and economic (commodities, prices) dynamics. However, models of the type presented here may be helpful in addressing dynamics of stocks, other than population or commodities. For example, variables like fear, greed, power, deviance, mass psychology; demographic, political, and social events, as, for instance, the leader–follower relationships, the bandwagon effect, over- and underreaction by individuals and collectives to social events, etc., could be effectively and productively modeled in their space–time dimensions along the lines suggested by our universal map. Of course, variables and events so diverse as these examples indicate, operate over different time–space constants. However, the formalisms employed to model their evolution may be quite similar and indeed may be subspecies of our universal relative map.

Dynamically violent events, possibly characterizing the evolution of these variables and the events alluded to earlier, are largely due to highly nonlinear connectance among closely linked variables and their interdependent (feedback) dynamic (iterative) structure. Dynamic models depicting spatial and temporal durability in these variables contain in their analytical form these stocks' generational (iterative) and spatial expansion–contraction (ticking) process.

A component central to the qualitative features of these models is the "complexity versus stability" issue, initially elaborated on by May (1974). According to May, as the dimensionality, strength, or degree of interconnectance among variables increases (i.e., complexity rises), then the likeliness that the systems will exhibit dynamically stable behavior diminishes. It is becoming increasingly apparent that no matter what the perceived dimensionality of socioeconomic systems or their perceived degree or strength of spatiotemporal connectivity, they are extensive and strong enough to exhibit very complex and unstable dynamic behavior. At times, their observed dynamical behavior is very similar to that which is characterized by bifurcation theory as "chaotic."

Simply deterministic, iterative processes capable of generating turbulent and chaotic phenomena challenge our perception that socioeconomic systems are stable and calm. These dynamic models open new windows into social evolution, as they indicate that elements of instability in social systems must be abundant and largely expected. Indeed, one must be surprised when the record shows stability and calmness in social events.

Analytically, the likelihood that dimensionally small and less disaggregated socioeconomic, spatiotemporal systems would demonstrate turbulent behavior may be lower than that of dimensionally large and highly disaggregated systems. As one shifts from one level of disaggregation to another, and as different regions of the dynamic model's parameter space are explored, different types of instability and turbulent behavior may emerge. Time scales, for such complicated social behavior to occur, vary among stocks, geographical locations, and chronological time periods.

An interesting phenomenon in spatiotemporal interdependence is that of "slaving" (Haken, 1983). According to this principle, stocks with longer time constants (frequencies) determine the behavior of stocks operating under shorter time constants. This is the, temporally deterministic, component of the slaving principle. However, in geographical systems, there is also a spatially deterministic component to the principle. Durability can be linked to areal size, so that stocks occupying smaller locations area-wise may be determined by stocks occupying larger areas. Spatially larger systems (e.g., national economies) may slave spatially smaller units within them (e.g., subnational, regional, or urban economies).

All these new concepts have resulted in numerous and exponentially growing publications in the natural sciences, pure and applied mathematics, mathematical ecology, and the socioeconomic spatial (geographic) sciences. Following the pioneering paper by Lorenz (1963) on turbulence in fluid dynamics, the Navier–Stokes flow equation, and the May (1976) discrete-time logistic prototype model, the literature (in many natural and social science fields) has grown very rapidly. It has reached the state where a comprehensive list of references is not feasible. The continuous and growing interest in the subject is evidenced in three related on-going series: H. Haken's series on *Synergetics*, published by Springer-Verlag, with examples of significant work found in Arnold and Lefever (eds.) (1981), Della Dora, Demongeot, and Lacolle (eds.) (1981), Eigen and Schuster (1979), Frehland (ed.) (1984), Haken (ed.) (1981), (1983), Schuster (1984), etc.; G. Nicolis and I. Prigogine's series on *Non Equilibrium Problems*, published by Wiley-Interscience, with examples of major contributions in, among others, Horton, Reichl, and Szebehely (eds.) (1983), Nicolis, Dewel, and Turner (eds.) (1981), etc.; and May's series in *Population Biology*, published by Princeton University Press. These series address natural science and the biological applications of nonlinear dynamics. Other significant work is that by Feigenbaum (1979), (1980), Gurel and Rossler (eds.) (1979), Helleman (1981), Sparrow (1982), Garrido (1983), Campbell, Crutchfield, Farmer, and Jen (1985), Holden (ed.) (1986), etc.

In the fields of socioeconomic sciences and geography, isolated works exist: in economics see, e.g., Barnett, Bernelt, and White (eds.) (1988) and an earlier paper by Day (1981); in geography, see Dendrinos (1984a), Dendrinos and Sonis (1987), (1988), Sonis and Dendrinos (1987a, b), (1988), and Reiner, Munz, Haag, and Weidlich (1986).

Periodicity and cycles in socioeconomic sciences have recently been analyzed, among others, by Casti (1984) and Nijkamp (1985), (1986), and by Dendrinos and Mullally (1981) and Dendrinos (with Mullally) (1985), in the spatial sciences (urban and regional science and geography). These works address numerous cycles of a daily, weekly, monthly, quarterly, annual, decennial, quarter century, and even longer-period, oscillatory motion in socioeconomic and demographic variables. In the area of economics, the business cycle has been addressed by, among others, Lucas (1981). Longer period cycles—the Kondratieff wave, the Schumpeterian clock, the Kuznets cycle, etc.—have been discussed, although not empirically documented by Nijkamp (1985), among others.

The recent discussion emerging from dynamical analysis, within the framework of the slaving principle mentioned earlier, highlights the central notions of "fast" and "slow" movements and their subcategories.

Various social events seemingly alter the socioeconomic systems spatiotemporal configuration through *fast* dynamics, which may include a phase transition. Environmental fluctuations, disturbances, and accidents, which could involve discontinuities or environmental catastrophes, act as shocks to the fast adjusting socioeconomic spatial dynamics. Sudden and sharp or continuous and smooth, topographical, climatic, or other changes associated with the natural environment are included in the environmental perturbations, which could induce changes in the fast socioeconomic spatiotemporal dynamics. These environmental fluctuations, frequently or infrequently occurring in sociological time, are assumed here to operate as *slow* dynamics. Their relative constancy is depicted by the current and held constant values of the dynamic model's parameters.

It would be of interest to supply a model with the maximum possible dynamic state variables' variability in its parameter space, so that the majority of these events could be depicted. The universal map we examine in this book, to our knowledge, contains the largest number of dynamic events among all known iterative maps in its parameter space.

Whether endogenously induced or exogenously influenced, phase transitions may occur in the spatiotemporal qualitative dynamics of socioeconomic stocks. In such cases the socio-spatial configuration is considered to have undergone *evolution.* These events are the result of the system having passed through critical points in its parameter space; in other words, the system has undergone a change in the qualitative properties of its underying socio-spatial dynamics. If, on the other hand, environmentally induced or not, dynamic changes in the state variable do not result in evolutionary change but merely in a smooth or abrupt continuation of current growth/decline paths, then *development* characterizes the current socio-spatial dynamics. Development is thus defined as slight shifts among or continuation of trajectories belonging to the currently prevailing equilibrium in the spatial dynamics of stocks. This topic is discussed at some length later in the text, in the context of the various slow and fast motions recordable in the universal discrete relative dynamics model.

2. *Some Simple Dynamic Models*

Interdependencies among socioeconomic stocks in space–time are complex and multifaceted, involving a bundle of forces at work. Within that bundle, social scientists have, at times, uncovered the necessary and sufficient linkages and central variables needed to replicate effectively and efficiently the observed behavior of socioeconomic systems.

In replicating, sufficiently close, the dynamic behavior of stocks distributed in space–time, certain simple (and rich in insight) iterative maps have been suggested. The simplest among them is the linear differential equation model

$$\frac{dx}{dt} = ax(t); \qquad x(t) > 0, \qquad -\infty \leq a \leq +\infty, \qquad \text{(I.A.2.1)}$$

identifying Malthusian exponential dynamics. The Verhulst–Pearl equation of logistic population (one-stock, one-location) growth

$$\frac{dx}{dt} = ax(t)[1 - x(t)], \qquad \text{(I.A.2.2)}$$

or the discrete-time, logistic prototype model (May, 1976)

$$x(t+1) = ax(t)[1 - x(t)]; \qquad 0 < a < 4, \quad 0 \leq x(t) \leq 1, \quad \text{(I.A.2.3)}$$

are examples of such simple but very interesting models of socio-spatial dynamics. All the complex and varied factors affecting the accumulation or decumulation of the stock are embedded in the parameter, *a*, and the form of the spatial or temporal interdependency with a time or space horizon. In these three models, one finds only one parameter (input) and one state variable (output). Thus, these are the most efficient continuous and discrete-dynamic models.

Another example of simple socio-spatial dynamics is found in the ecological Volterra–Lotka model of population interactions

$$\frac{dx_i}{dt} = x_i\left(a_i + \frac{1}{b_i}\sum_j a_{ji}x_j\right), \qquad i = 1, 2, \ldots, I. \tag{I.A.2.4}$$

In this case, the population growth rate of a species is a linear function of the abundance of all species. The model was initially proposed as a nonspatial community of an I-species interaction model in mathematical ecology. Here, a_i is the "self-growth" rate, b_i is an "average weight" of the species, and the coefficients a_{ji} are the interaction parameters. It is noted that the efficiency of this formalism is lower than in the Malthusian case, since now there are I state variables and $I(2 + I)$ parameters. However not all of these parameters need be nonzero. Dynamic properties of this model have been shown to depend on the number of interacting species. The model is also capable of producing complex behavior, even in the case of a predator–prey model (Gilpin, 1979).

Since the model's behavior depends crucially on the numbr of state variables, a central result in socioeconomic space–time dynamics emerges: one cannot arbitrarily subdivide populations (i.e., spatially or otherwise disaggregate the model at will), keep the time–space horizon and the dynamic formalism the same, and at the same time retain the desired dynamic features. It must be expected that there is a level of disaggregation, a space framework or a time horizon, for which such a formalism optimally holds. For, below or above it, and in shorter or longer time periods, the introduction of new social forces necessitates that the formalism itself must be modified. Central qualitative properties of the dynamics obtained by a specific breakdown of such a formalism do not remain intact, as finer or more coarse levels of disaggregtion are considered. This issue will be addressed further in subsequent sections of this book.

By taking such an approach to socio-spatial systems dynamics one is forced to answer basic questions like: What is the appropriate socioeconomic disaggregation of the central state variables? (a question related to the definition of homogeneity of any social stock, e.g., population); How many central variables are involved in socioeconomic spatial dynamics? (a question related to the efficient socio-spatial "problem definition" and the choice of appropriate interdependencies among the "core" state variable(s) and parameters and their peripheral counterparts); What is the appropriate spatial disaggregation? (a question related to space homogeneity); and What is the appropriate way to formulate dynamic interdependency? (a question related to time homogeneity). Finally, the question is raised as to whether to model dynamics either through discrete time (iterations) or continuous dynamics; the decision implies that one must account for "generational" aspects in the state variable(s). Also associated with it is a decision on whether to model in either continuous or discrete space. These questions touch deeper issues of social epistemology, far beyond the scope of this book.

3. *The Location-Dependent Elements of a Theory of Socio-Spatial Dynamics*

In this subsection we provide a brief outline of the central components for constructing a general theory of socio-spatial evolution. It sets the stage for presenting the theoretical models which follow. Let us start by stating in abstract that a multiplicity of social stocks are present at any point in space–time. Each homogeneous social stock is distributed in heterogeneous discrete space extended over a spatial horizon I, so that a closed environment of I locations is considered to be the theater of the socioeconomic forces considered. Within this spatial boundary, and at any time period contained within a time horizon T, a variety of socioeconomic stocks interact. Examples of such interstock heterogeneous, intrastock homogeneous quantities are the population, built capital, economic output, information, and other stocks found at a particular point in space–time. Various natural resources are also examples of nonuniformly distributed stocks in space–time.

Each location's stocks are characterized, due to topographical and other geographic factors by differential access to other locations' stocks, at any time period. Differential access to natural resources, and to spatially distributed socioeconomic production and consumption points and markets, renders space intrinsically heterogeneous.

The bundle of elements behind space heterogeneity constitutes composite *locational advantages* enjoyed at a particular point in space–time. When comparing the locational advantages among locations, composite *comparative advantages* are obtained. Comparative advantages are interrelated with the observed spatiotemporal distributions of stocks. Comparative advantages are functions of socioeconomic, spatiotemporal distributions. In turn, these socioeconomic, spatiotemporal distributions are functions of comparative advantages. One does not exist in the absence of the others. Causes and effects vanish within this interrelationship.

B. The Four Lenses to View Socio-Spatial Dynamics

1. *The Absolute–Relative Lens*

Two of the many possible lenses for viewing socio-spatial dynamics are the absolute and relative lenses. Two other lenses are the continuous and discrete modes. Combining the two, one can have four alternative modes to model socio-spatial dynamics. The use of an absolute lens implies that the observer considers an unbounded environment within which *open* and, subject to fluctuations emanating from the environment, locally interconnected systems interact. On the other hand, use of a relative lens implies that the observer consider a *closed* environment within which local systems interact. The nature of fluctuations in the last case is not immediately apparent.

Under the absolute lens there are non-zero-sum games played out among locations and stocks; whereas, under the relative lens, there are only zero-sum games played out among either various locations and/or stocks. The gap in the literature on relative (purely competitive) socio-spatial dynamics is much wider than of absolute dynamics. Absolute dynamics are appropriate to analyze open systems, where the environment is not well defined. Relative dynamics are used for examining closed systems where the environment is well defined—the environment being the area over which one normalizes the various stocks' size (Dendrinos (with Mullally), 1985). The work here almost exclusively elaborates on relative discrete dynamics over space.

Drawing from well-established formalizations from our neighboring field, mathematical ecology, and from previous geographic–regional science tradition, the overwhelming majority of the newly proposed spatial models address *absolute* change in their state variables. However, recently identified theoretical concerns, in connection with the available empirical evidence, created a great deal of interest in the development of nonlinear models of *relative* change. Central to these latest models is the dynamics of relative distribution over space of a homogeneous stock (e.g., population, capital, etc.).

These models can address a variety of social science topics. A list of examples in the spatial social sciences (geography, and urban/regional economics) includes: issues related to the dynamics of various socioeconomic and spatial distributions, identifying spatial or sectoral disparities or dualisms; spatio-temporal market processes (decentralized decision-making) involving monopolies (monopsonies), oligopolies (oligopsonies); inter- and intraurban, inter- and intraregional relative growth and decline of many industrial sectors; multiregional or multinational trade and competition; population survival or extinction; various cycles of economic activity; diffusion of technological and policy innovations, etc. A few examples of such relative dynamics are found in Dendrinos (with Mullally) (1985), Sonis (1983a, b), (1984), (1985), (1986a, b), (1987), among others.

A major element of the nonlinear models of relative dynamics is the existence of some specific conservation conditions, which hold at all time periods and constrain the dynamic paths of such systems. These conditions differ from the postulated conservation laws found in earlier work by Volterra (1927) on absolute growth. A detailed analysis of two conservative systems' continuous dynamics and their potentials is found in Dendrinos and Sonis (1986). The paper derived a governing integral of a *cumulative temporal entropy* giving rise to these dynamics interpreted as a cumulative socio-spatial utility. It was viewed as an aggregate comprehensive governing principle in social systems.

a. The Specifications

In this section we provide the general background for the multidimensional (many-location), one-stock absolute and relative dynamics model, and expose its basic properties. Further, we supply a process for generating and classifying various specifications of relative dynamics. Their analytical properties are demonstrated by an attempt to derive a classification of the universal relative dynamics model's stability conditions, to be discussed more fully in subsections which follow.

Consider a homogeneous substance (e.g., human population stock) distributed over a nonhomogeneous space—for the moment assumed to be arbitrarily subdivided into I regions. The *absolute behavior* of the system is fully described at any time period t by the spatial absolute population levels of the I regions, defining the socio-spatial configuration of this one-stock case at any time period. The community of regions' vector of state variables $\mathbf{X}$ defines at t the current population stock distribution

$$\mathbf{X}(t) = [X_1(t), X_2(t), \ldots, X_I(t)], \qquad \text{(I.B.1.1)}$$

$$X_i(t) > 0, \qquad i = 1, 2, \ldots, I, \qquad \text{(I.B.1.2)}$$

the elements of which, summed up in turn, define the total level of the homogeneous stock in the community of the I regions

$$X(t) = \sum_{i=1}^{I} X_i(t) > 0. \qquad \text{(I.B.1.3)}$$

From the above, a fundamental requirement in the definition of spatial heterogeneity emerges: in order for a region i^* to be a member of the set of regions included in the spatial disaggregation into I regions of the (spatially fixed) community, it must contain a nonzero quantity of the homogeneous stock (e.g., population) at least in one time period t in either the discrete- or continuous-time horizon interval ($t = 1, 2, \ldots, T$ or $t \in T$).

b. Relative Dynamics

The *relative behavior* of the system is fully described at t by the spatial relative population levels of the I regions, defining a probabilistic vector $\mathbf{x}(t)$

$$\mathbf{x}(t) = [x_1(t), x_2(t), \ldots, x_I(t)], \qquad \text{(I.B.1.4)}$$

$$x_i(t) > 0, \qquad i = 1, 2, \ldots, I \quad (t = 1, 2, \ldots, T \text{ or } t \in T), \qquad \text{(I.B.1.5)}$$

satisfying the identity

$$x_i(t) = \frac{X_i(t)}{\sum_{j=1}^{I} X_j(t)} = \frac{X_i(t)}{X(t)}; \qquad i = 1, 2, \ldots, I \quad (t = 1, 2, \ldots, T \text{ or } t \in T). \qquad \text{(I.B.1.6)}$$

Note that

$$\sum_{i=1}^{I} x_i(t) = 1, \tag{I.B.1.7}$$

$$x_i(t) = \frac{X_i(t)}{\sum_{j=1}^{I} X_j(t)}, \quad i = 1, 2, \ldots, I \quad (t = 1, 2, \ldots, T \text{ or } t \in T). \tag{I.B.1.8}$$

The system's dynamics when deterministically defined require some initial perturbation, given by the starting values vector $\mathbf{X}(0)$, and a set of dynamical equations in either discrete or continuous fashion.

The system we are about to analyze in this one-stock example is "singularly" interdependent. Changes in the population size (or any homogeneous stock we consider distributed in space), in either discrete- or continuous-time, depend only on the current population level. There are no other state variables. If that is so, then the system is *temporally interdependent* over one state variable, in this case, population. There may be more than one type of state variable in the system; such "multiply interdependent" cases will be analyzed later in a broad manner. However, model efficiency is sought, such that the necessary and sufficient parameters and variables are contained for describing the dynamics of the system, in view of a particular observed dynamical behavior.

2. *Continuous Dynamics*

In this case, some general results can be obtained relating relative, absolute, and total community dynamics regardless of the specifications for growth. From the relative size identity (I.B.1.6) one obtains either of the following conditions (dropping the subscript t)

$$\frac{\dot{x}_i}{x_i} = \frac{\dot{X}_i}{X_i} - \frac{\dot{X}}{X}, \quad i = 1, 2, \ldots, I,$$

$$\frac{\dot{x}_i}{x_i} - \frac{\dot{x}_k}{x_k} = \frac{\dot{X}_i}{X_i} - \frac{\dot{X}_k}{X_k}, \quad i, k \in I, \tag{I.B.2.1}$$

or, introducing the elasticity of growth ε_i,

$$\varepsilon_i = \frac{\dot{X}}{X} \bigg/ \frac{\dot{X}_i}{X_i}, \quad i = 1, 2, \ldots, I, \tag{I.B.2.2}$$

$$\frac{\dot{x}_i}{x_i} = \frac{\dot{X}_i}{X_i}(1 - \varepsilon_i), \quad i = 1, 2, \ldots, I, \tag{I.B.2.3}$$

$$\dot{x}_i = \frac{\dot{X}_i}{X}(1 - \varepsilon_i), \quad i = 1, 2, \ldots, I,$$

$$\frac{\dot{x}_i}{\dot{x}_k} = \frac{\dot{X}_i}{\dot{X}_k} \frac{(1 - \varepsilon_i)}{(1 - \varepsilon_k)}, \quad i, k = 1, 2, \ldots, I. \tag{I.B.2.4}$$

From (I.B.2.1) the first basic relationship on relative dynamics is derived: the relative regional growth rate is equal to the absolute regional growth rate minus the total community growth rate. From (I.B.2.3) a corollary is derived: in positive absolute regional growth rate conditions, the relative regional growth rate is positive/negative if the absolute growth elasticity with respect to the community's total growth rate is less/more than unity.

The continuous relative regional growth rate is *not*, in the general case, continuously increasing/decreasing. This can be checked directly from the following

$$\frac{\partial}{\partial X}\left(\frac{\dot{x}_i}{x_i}\right) = \frac{\dot{X}}{X^2} - \frac{1}{X}\frac{\partial \dot{X}}{\partial X} \gtrless 0, \tag{I.B.2.5}$$

$$\frac{\partial}{\partial X_i}\left(\frac{\dot{x}_i}{x_i}\right) = \frac{1}{X_i}\frac{\partial \dot{X}_i}{\partial X_i} - \frac{1}{X}\frac{\partial \dot{X}}{\partial X_i} + \frac{\dot{X}}{X^2} - \frac{\dot{X}_i}{X_i^2} \gtrless 0. \tag{I.B.2.6}$$

In the continuous case the problem is fully defined by considering vector functions relating X_i and X in an interdependent manner

$$\begin{aligned} \dot{\mathbf{X}}(t) &= \mathbf{F}[X(t), \mathbf{P}] = 0, \\ \mathbf{X}(0) &> 0, \end{aligned} \tag{I.B.2.7}$$

where F_i, $i = 1, 2, \ldots, I$, identify functions of the core state variables vector $\mathbf{X}$ and other variables and/or parameters $\mathbf{P}$.

Studies, using Volterra–Lotka relative growth specificatons for type (I.B.2.1) dynamics, conclude that the right-hand side of the above continuous dynamics can be modeled well by a linear function of relative urban size in a community of U.S. metropolitan areas (Dendrinos (with Mullally), 1985). This in turn implies

$$\begin{aligned} \dot{x}_i &= x_i F_i, \\ F_i &= a_i - b_i x_i, \qquad a_i, b_i > 0, \end{aligned} \tag{I.B.2.8}$$

or

$$\dot{x}_i = b_i x_i (c_i - x_i), \qquad c_i = \frac{a_i}{b_i}, \tag{I.B.2.9}$$

a logistic-type growth model with a carrying capacity c_i. Linear F_i in (I.B.2.8) implies, due to (I.B.2.1),

$$\frac{\dot{X}_i}{X_i} = F_i + \frac{\dot{X}}{X} = a_i + \frac{\dot{X} - b_i X_i}{X}, \tag{I.B.2.10}$$

indicating that the absolute growth rate of U.S. metropolitan areas is given by a constant (but subscripted by area) parameter, plus a fraction of the national growth rate diminished by size effects (externalities of agglomeration picked up by the size of parameter b_i).

Let us analyze a specific case where in continuous dynamics the absolute regional growth functions are specified by *exponential* growth/decline conditions, so that

$$\dot{X}_i = a_i X_i,$$
$$X_i(t) = X_i(0)\exp(a_i t), \qquad \text{(I.B.2.11)}$$
$$-\infty \le a_i \le +\infty, \qquad X_i(0) > 0, \qquad i = 1, 2, \ldots, I,$$

with a bifurcation in behavior when $a = 0$. Specification (I.B.2.11) is of interest because it represents regional Malthusian growth, a model often closely approximating phenomena widely observed in real life in a broad spectrum of regional population systems and over limited, as well as extended, time horizons.

The important fact is that absolute Malthusian growth generates *logistic growth* relative regional dynamics

$$x_i = \frac{X_i(0)\exp(a_i t)}{\sum_{j=1}^{I} X_j(0)\exp(a_j t)}$$
$$= 1/1 + \sum_{\substack{j\neq i\\ j=1}}^{I} \left[\frac{X_j(0)}{X_i(0)}\right] \exp[(a_j - a_i)t], \qquad 0 < x_i < 1, \quad i = 1, 2, \ldots, I. \qquad \text{(I.B.2.12)}$$

The quantities x_i are not defined at either limit, and they always satisfy the conservation condition

$$\sum_{i=1}^{I} x_i = 1. \qquad \text{(I.B.2.13)}$$

Note that these quantities may approach, abruptly or asymptotically, zero or one. Thus, in multiregion relative dynamics, extinction may be asymptotically approached, so that it represents a stable steady state the system converges toward, or the result of unstable behavior. Note that the relative size condition can also be written as

$$\frac{1 - x_i}{x_i} = \sum_{\substack{j\neq i\\ j=1}}^{I} \left[\frac{X_j(0)}{X_i(0)}\right] \exp[(a_j - a_i)t]. \qquad \text{(I.B.2.14)}$$

Define as $x_i^-(t)$ the *antiregional* size of region i, given by

$$x_i^-(t) = 1 - x_i(t) = \sum_{\substack{j\neq i\\ j=1}}^{I} x_i(t); \qquad \dot{x}_i(t) = -\dot{x}_i^-(t), \qquad \text{(I.B.2.15)}$$

then

$$x_I(t) = 1/1 + \sum_{j=1}^{I-1} \left[\frac{X_j(0)}{X_I(0)}\right] \exp[(a_j - a_I)t]$$
$$= 1 - x_I^-(t) = 1 - \sum_{k=1}^{I-1} \left\{ 1/1 + \sum_{\substack{j\neq k\\ j=1}}^{I} \left[\frac{X_j(0)}{X_k(0)}\right] \exp[(a_j - a_k)t] \right\}, \qquad \text{(I.B.2.16)}$$

for any "numeraire" region being designated as I. The notion of "antiregion" was first developed by Dendrinos (1984b). Sometimes it is helpful to assume some direct (approximate) relationship between $x_I^-(t)$ and $x_I(t, t-1, \ldots)$, instead of specifying interactions among all locations/stocks as shown above. The point is to identify the complex events connecting $x_I(t)$ and $x_I^-(t)$ without resorting to the complicated formulations underlying the behavior of *each* region, within a community of regions, if one wishes to window into the behavior of one of them. If, over a time horizon T, one can identify a non-random relationship between x_i and x_i^-, then the notion of an "antiregion" is useful. From (I.B.1.12), differentiating with respect to t one obtains

$$\dot{x}_i = x_i\left[\frac{a_i - \sum_{j=1}^{I} a_j X_j(0)\exp(a_j t)}{\sum_{j=1}^{I} X_j(0)\exp(a_j t)}\right]$$

$$= x_i\left(a_i - \sum_{j=1}^{I} a_j x_j\right) = x_i \sum_{j=1}^{I} (a_i - a_j)x_j. \qquad \text{(I.B.2.17)}$$

From this condition one further derives

$$\frac{\dot{x}_i}{x_i} = a_i - \sum_{j=1}^{I} a_j x_j \qquad \text{(I.B.2.18)}$$

and for any two regions

$$\frac{\dot{x}_i}{x_i} - \frac{\dot{x}_k}{x_k} = \frac{\dot{x}_i}{X_i} - \frac{\dot{x}_k}{X_k} = a_i - a_k. \qquad \text{(I.B.2.19)}$$

Introducing the coefficients

$$a_{ij} = a_i - a_j \qquad \text{(I.B.2.20)}$$

one has

$$\dot{x}_i = x_i\left(\sum_{j=1}^{I} a_{ij} x_j\right), \qquad \text{(I.B.2.21)}$$

which is the Volterra–Lotka model of population interactions (I.A.4.4) with zero self-growth. This system of nonlinear differential equations has also appeared in the Innovation Diffusion Theory to describe the spatial spreading of totally antagonistic competitive innovations (Sonis, 1983a, b). A more general system, with arbitrary interaction coefficients a_{ij}, represents the diffusion of competitive innovations participating in cooperative zero-sum games.

System (I.B.2.21) represents the general case of interregional competition or interregional dynamics of relative population stocks. The community matrix $A = (a_{ij})$ describes the effect of each region upon the relative growth of all other regions. The important fact is that, due to the conservation condition (I.B.2.13), for the system to be internally consistent the community matrix $A = (a_{ij})$ must be antisymmetric (Sonis, 1983a, pp. 103–104)

$$a_{ij} + a_{ji} = 0, \quad a_{ii} = 0. \qquad \text{(I.B.2.22)}$$

One can interpret this antisymmetry as follows: each pair, i, j, of regions participates in an antagonistic zero-sum game where the quantity a_{ij} is an expectation of composite net gain from the interaction between the ith and the jth regions.

Such an interpretation implies that *competitive exclusion* among regions may occur: if the ith region is a winner in all antagonistic games against all other regions, i.e.,

$$a_{ij} > 0 \quad \text{for all} \quad j \neq i, \tag{I.B.2.23}$$

then in the long run the total quantity of the population stock expressed in relative terms will be transferred and concentrated within the winning region. These heuristic hints are based on elementary linear stability analysis (Sonis 1983a; Dendrinos and Sonis, 1986): the competitive exclusion states of the form $\mathbf{x} = (0, 0, \ldots, 0, 1, 0, \ldots, 0)$ are asymptotically stable under the conditions (I.B.2.22, 23).

One can formulate the entropy measure of the relative size population association in a community of regions at any point in time

$$H = -\sum_{i=1}^{I} x_i \ln x_i = \sum_{i=1}^{I} x_i \ln \frac{1}{x_i}. \tag{I.B.2.24}$$

Competitive exclusion dynamic equilibria produce a minimum in the entropy H; in other words, *the competitive exclusion principle in relative dynamics is equivalent to a principle of dynamic entropy minimization.* A most important finding in spatiotemporal population interaction is the derivation of the relative spatial dynamics (I.B.2.21) from Hamilton's variational principle (Dendrinos and Sonis, 1986). Let us consider the cumulative portions of relative population

$$Y_i(t) = \int_0^t x_i(t)\, dt, \qquad \dot{Y}_i(t) = x_i(t). \tag{I.B.2.25}$$

It is possible to choose the variational integral

$$E = \int_0^T \left(-2 \sum_i x_i \ln x_i + \sum_i \sum_j a_{ij} x_i Y_j \right) dt, \tag{I.B.2.26}$$

the integrand of which includes the entropy of the spatial distribution of relative population and the interaction among portions of relative population allocated to different regions. The analogue of Hamilton's principle of stationary action means that the first variation of the integral E vanishes, giving rise to the system of Euler differential equations. For the case of the integral (I.B.2.26), the Euler conditions coincide with the relative spatial dynamics (I.B.2.21).

The stationary value of the integral E turns out to be the temporal cumulative entropy

$$\int_0^T \left(-\sum_i x_i \ln x_i \right) dt = \int_0^T H\, dt \tag{I.B.2.27}$$

for the relative spatial dynamics over the time horizon T. This is of particular interest to socio-spatial dynamics, since entropy over space evolves to a minimum in the asymptotically stable competitive exclusion equilibrium state. Over time, the governing integral (I.B.2.26), which includes the accumulation of spatial entropy and interactions, merges into a maximum, the stationary value of temporal cumulative entropy. One may ponder nondirect optimal control and aggregate social welfare aspects of this variational principle, a subject more fully addressed in the book by Dendrinos (with Mullally) (1985) and the paper by Dendrinos and Sonis (1986).

An interesting extension of the above constant growth rate absolute regional dynamics is the set of specifications

$$\frac{\dot{X}_i}{X_i} = \alpha_i - \beta_i X_i, \tag{I.B.2.28}$$

implying that friction is present in the association. Still another interesting extension might be the specifications

$$\dot{X}_i = \alpha_i X_i \pm \beta_i X_i^2 \pm \gamma_i X_i^3 \pm \delta_i X_i^4 + \cdots; \quad \alpha_i > \beta_i > \gamma_i > \delta_i \ldots, \tag{I.B.2.29}$$

whereby growth rates in absolute regional population levels are affected by factors proxied by higher-order terms of absolute population size. This concludes our analysis of the highlights found in the continuous dynamics case.

It is of interest to note finally that a multistock and multilocation Volterra–Lotka-type relative continuous dynamics model is degenerate (Sonis and Dendrinos, 1988). That is, the form

$$\dot{x}_{ij}(t) = \left[a_{ij}^0 + \sum_k \sum_m a_{ij}^{km} x_{ij}^{km}(t) \right] x_{ij}(t), \tag{I.B.2.30}$$

where the indices i and k designate location, and j and m stand for stock type, collapses to

$$\dot{x}_{ij}(t) = \left[a_{ij}^0 + \sum_k a_{ij}^k x_{ij}^k(t) \right] x_{ij}(t), \tag{I.B.2.31}$$

so that the original system corresponds to J isolated dynamic models: each stock type at a location interacts with similar stocks at all other locations! This justifies why Volterra and Lotka analyzed isolated communities of many species: our result indicates that Volterra–Lotka-type dynamics are nondegenerate, either for interactions among one stock in many locations or among many stocks in one location.

3. *Discrete Dynamics: The Universal Discrete Relative Dynamics Model*

In the discrete case which contains a temporal interdependency among states of a multiregional one-population stock the system obtains the form, in vector notation

$$\begin{aligned} &\mathbf{X}(t+1) = \mathbf{F}[\mathbf{X}(t)] > 0, \quad t = 0, 1, 2, \ldots, \\ &\mathbf{X}(0) > 0. \end{aligned} \tag{I.B.3.1}$$

The continuous case represents dynamics with an overlap among "generations", whereas, in the discrete case, there is no overlap between two successive generations, one influencing the level of the other with a time lag of one or more time periods. Generational overlaps may hold in the case of *absolute* population growth. In the case of *relative* population dynamics, generational overlap is meaningless. Instead, a succession may be taking place in either a continuous or a discontinuous manner in the size of different elasticities of the local stock growth with respect to the environment's growth.

Consider a set of arbitrary positive real-valued functions (vector function $\mathbf{F} = (F_1, F_2, \ldots, F_I)$) such that each F_i is defined at each time period t by a subset of x_i's in $\mathbf{x}$, as long as at least one of the F_i's depends on at least one of the x_i's, a necessary condition for dynamical behavior. We will call this condition the "inclusion principle." In the exposition to follow, and without loss of generality, we will assume

$$\begin{aligned} &F_i[x_1(0), x_2(0), \ldots, x_I(0)] > 0, \\ &0 < x_i(0) < 1, \quad i = 1, 2, \ldots, I, \\ &\sum_{i=1}^{I} x_i(0) = 1, \\ &F_I = 1, \end{aligned} \tag{I.B.3.2}$$

where the x_i's are not defined, in the general case, at the limits. Define the iterative process

$$x_i(t+1) = \frac{F_i[\mathbf{x}(t)]}{\sum_{j=1}^{I} F_j[\mathbf{x}(t)]}, \quad i = 1, 2, \ldots, I, \quad t = 1, 2, \ldots, T \tag{I.B.3.3}$$

or a more-than-one-time-period delay interdependency

$$x_i(t+n) = \frac{F_i^n[\mathbf{x}(t)]}{\sum_{j=1}^{I} F_j^n[\mathbf{x}(t)]}, \quad i = 1, 2, \ldots, I, \quad t = 1, 2, \ldots, T \tag{I.B.3.4}$$

where F_i^n is the nth iterate of F_i (F_i composed with itself n times), and where n is an integer varying between $(1, +\infty)$. The n-time delay discrete-iterative case would need n initial conditions, so that

$$\sum_{i=1}^{I} x_i(0) = 1, \sum_{i=1}^{I} x_i(1) = 1, \ldots, \sum_{i=1}^{I} x_i(n-1) = 1. \tag{I.B.3.5}$$

This extension will not be pursued further here, as we will always assume that

$n = 1$. From the above, one can easily deduce that

$$x_i(t+1) = 1 \bigg/ \left\{ 1 + \sum_{\substack{j \neq i \\ j=1}}^{I} \frac{F_j[\mathbf{x}(t)]}{F_i[\mathbf{x}(t)]} \right\}, \qquad i = 1, 2, \ldots, I,$$
$$0 < x_i(t+1) < 1, \qquad i = 1, 2, \ldots, I, \qquad \text{(I.B.3.6)}$$
$$\sum_{i=1}^{I} x_i(t) = 1, \qquad t = 1, 2, \ldots .$$

Thus, this model automatically generates the temporal sequence of probabilistic vectors $\mathbf{x}(t)$ for all $t \in T$. Condition (I.B.3.6) is a *universal* discrete-time model of relative dynamics, since it contains any explicit form of the (arbitrary) vector function $\mathbf{F}$ satisfying the "inclusion principle." It enables one to generalize the results shown below (in all their variety, which includes turbulence) to systems which have functions $\mathbf{F}$ defined over any other (nonpopulation) variables, as long as *at least one* $\mathbf{F}$ is population dependent.

The universality of discrete-time relative dynamics (I.B.3.3) is the direct implication of the fact that given any specifications such that the inclusion principle is met and

$$x_i(t+1) = F_i[\mathbf{x}(t)], \qquad i = 1, 2, \ldots, I,$$
$$0 < x_i(t) < 1, \qquad i = 1, 2, \ldots, I, \qquad \text{(I.B.3.7)}$$
$$\sum_{i=1}^{I} x_i(t) = 1,$$

it directly follows that

$$F_i[\mathbf{x}(t)] > 0, \qquad i = 1, 2, \ldots, I,$$
$$\sum_{i=1}^{I} F_i[\mathbf{x}(t)] = 1, \qquad \text{(I.B.3.8)}$$

so that

$$x_i(t+1) = \frac{F_i[\mathbf{x}(t)]}{\sum_{j=1}^{I} F_j[\mathbf{x}(t)]}, \qquad i = 1, 2, \ldots, I, \quad t = 1, 2, \ldots, T. \qquad \text{(I.B.3.9)}$$

A two-variable, simple model of the above discrete-time dynamics is the well-known (May, 1976) fixed-point mapping

$$x(t+1) = ax(t)[1 - x(t)], \qquad 0 \leq x \leq 1, \quad 0 \leq a \leq 4. \qquad \text{(I.B.3.10)}$$

After the transformation

$$x(t+1) = 1/1 + \frac{1 - ax(t)[1 - x(t)]}{ax(t)[1 - x(t)]}$$

this model can be put into the universal form through the substitutions

$$x_1(t+1) = \frac{F_1}{F_1 + F_2[x_1(t), x_2(t)]},$$
$$x_2(t+1) = F_2[x_1(t), x_2(t)],$$
$$F_1 = 1, \tag{I.B.3.11}$$
$$F_2[x_1(t), x_2(t)] = \frac{1 - ax_1(t)x_2(t)}{ax_1(t)x_2(t)} > 0,$$
$$x_1(t) + x_2(t) = 1.$$

May's restriction that the parameter, a, varies between zero and four results from the positivity requirements imposed on F_1 and F_2; in this case, the $F_2 > 0$ constraint requires that

$$F_2[x_1(t), x_2(t)] > 0; \qquad 0 < ax_1(t), \quad [1 - x_1(t)] < 1. \tag{I.B.3.12}$$

It follows that for $x_1(t+1)$ not to exceed one or fall below zero (note that the maximum of $x_1(t+1)$ is equal to $a/4$ at $x_1(t) = 0.5$), upper and lower bounds must hold for the parameter, a, at $a_{min} = 0$ and $a_{max} = 4$.

The May example, extensively analyzed in the literature, demonstrates only a small part of the extraordinarily rich behavior of the universal discrete-time relative dynamics. General dynamic proprties of the model include features found in other discrete maps, like the quasi-periodicity event found in the Curry–Yorke (1978) map. These similarities and events are to be explored in forthcoming work.

The introduction of arbitrary parameters into the iterative process through $\mathbf{F}[\mathbf{x}(t)]$ leads to the study of conditions under which the qualitative features (stability) of the nonlinear interdependencies are preserved. Or, to conditions where structural change emerges through transitions from one kind of dynamic spatial stock distribution path to another.

Part II deals exclusively with the case of one-homogeneous stock (population), two-region, discrete-time dynamics. Even in this simple case, the study of dynamic equilibria, period-doubling cycles, and deterministic chaos cannot always be obtained in closed form solutions (analytically). Computer simulation is required. But in the most interesting case of the log-linear specification of the F functions

$$F_1 = A_1 x_1(t)^{\alpha_1} x_2(t)^{\alpha_2}, \qquad F_2 = A_2 x_2(t)^{\beta_1} x_2(t)^{\beta_2}, \tag{I.B.3.13}$$

a number of key analytical results can be obtained. This specification is of interest because it contains all possible associations among stocks and locations, in the sign of its real exponents.

The present study exhausts the analytical treatment for the universal iterative process, by focusing on some very broad and interesting specifications of pertinence to regional population distribution analysis. An analytical and an associated geometric procedure are set up to construct and identify the stability domains of the equilibrium states for the one-stock, two-region discrete dynamics case. They provide a heuristic basis for computer simulation and for numerical search when analysis is limited.

There are vast areas in the parameters' space where one can look for very particular events occurring in extremely narrow bands in that space, not possible to detect analytically. In the neighborhood of critical points in the parameter space where thresholds are crossed, unknown until now and known events occur in the state variables' space, including various phase transitions on the road to chaos. Such events can be detected through computer simulation, starting from relatively short distances in the parameters' space to these thresholds. One must be very judicious in the way resources are spent in examining the most promising areas of the parameters' space, and in detecting and recognizing novel dynamical phenomena. Computer simulation results reported in the following parts capture only certain of the many events embedded in the universal map. Future research, no doubt, will unravel many more.

Conclusions

In this part we introduced our universal discrete relative map and placed it in its analytical and mathematical perspective. Briefly discussed was the one-stock, two-locaton case relating this universal algorithm to May's logistic prototype. A number of examples were provided in an effort to outline the social sciences and the geographic applications of our mapping. Spatio-temporal interdependencies of stocks, and periodicity in socio-spatial dynamics, were presented as the central elements for laying down the foundations of a universal map of socio-spatial evolution formed within a relative, discrete in time–space, framework.

Part II
One Stock, Two Regions

Summary

In Section A of this part we first consider some general analytical properties of the one-stock, two-region, discrete-time map, which is a special case of our universal discrete relative dynamics mapping. This section focuses on the stability properties of the fixed point. We analyze its *first* iterate and supply a few comments on its higher ones. A corresponding geometric algorithm, used for analyzing the behavior of equilibria, is elaborated on. We then proceed to examine the broader ecological phenomenon of competitive exclusion within the framework of this map and provide a classification scheme. In Section B the one-stock, two-location, log-linear model is presented, highlighted by some special cases at particular points of interest in the parameter space.

Section C addresses *higher* iterates; the discrete equivalent of the Hopf bifurcation in continuous dynamics, period-doubling behavior, and Feigenbaum sequences are described. Deterministic chaos and transitions from one type of turbulence to another are also reviewed.

Finally, Section D examines in brief an *exponential* specification of the universal map. Throughout the exposition analytical results are supplemented with numerical findings, although formal proofs are not given.

A. The First Iterate and Associated Analytical Properties of the Model: $x(t + 1) = 1/1 + AF[x(t)]$

1. *Overview*

A new map in the mathematical literature of fixed-point algorithms is introduced, in an attempt to depict temporal interdependencies of socioeconomic stock(s) distributed over space, in relative terms. In subsection 2 we present its fixed-point behavior and stability properties, as we analyze the general case (i.e., nonspecific F functions). A new general geometrical algorithm is introduced to analyze the dynamics of equilibria. The behavior of the successive iterates of the discrete map is also sketched. In subsection 3 we focus on competitive exclusion configurations; whereas in subsection 4 we address a general classification scheme for such a family of algorithms.

2. *Fixed-Point Behavior and the Discrete Map*

a. The Specifications

Let us consider the following one-stock, two-location version of a discrete-time, discrete-space, one-time-period-delay, relative, nonlinear map

$$x_1(t+1) = \frac{A_1 F_1[x_1(t), x_2(t)]}{W}, \tag{II.A.2.1}$$

$$x_2(t+1) = \frac{A_2 F_2[x_1(t), x_2(t)]}{W}, \tag{II.A.2.2}$$

$$W = A_1 F_1 + A_2 F_2, \tag{II.A.2.3}$$

$$A_1, A_2 > 0,$$

$$F_1, F_2 > 0,$$

$$0 < x_1(t), x_2(t) < 1, \qquad t = 1, 2, \ldots, T,$$

$$x_1(t) + x_2(t) = 1.$$

In the above fixed-point algorithm, the next time period stock abundance at any location is a directly proportional function of the current locational and temporal advantages, F, found there and then, and inversely proportional to the current advantages encountered in the other location competing for that stock. Temporal locational advantages, depicted by the F_1 and F_2 functions, are smooth and strictly positive. They are, in turn, dependent upon the current abundance of the stock at the two locations, $x_1(t)$ and $x_2(t)$. Two speeds of change are built into this map; x_1, x_2 (the state variables) experience *fast* movement when compared with relatively *slow* changes in the environmental parameters $[A]$.

Thus, an interdependence is built into this model so that the size of the stock in the two locations is the only state variable affecting the stock's dynamics along the Malthusian tradition. Possible interactions of the stock's abundance in the two locations with other stocks are built into the mapping through the model's parameters (A_1, A_2 and those found in the specifications of the F_1, F_2 functions).

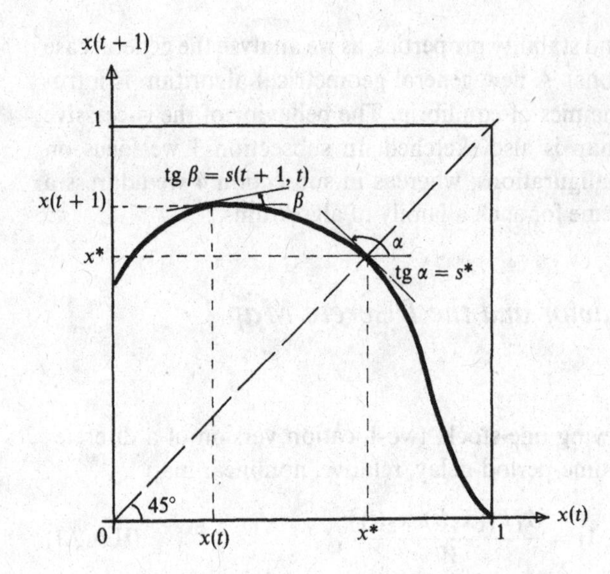

FIGURE 1. The first iterate of the fixed-point process $x(t+1) = 1/1 + AF[x(t)]$; $s(t + 1, t)$ is the slope of the mapping at any point $x_1(t)$, whereas s^* is the slope at equilibrium.

Note that specifications (II.A.2.1–3) can also be written as

$$x_1(t + 1) = 1/1 + AF_0,$$
$$x_2(t + 1) = 1 - x_1(t + 1), \tag{II.A.2.4}$$

where $A = A_2/A_1 > 0$ and $F_0 = F_2/F_1 > 0$. Function $F_0 = F_0[x_1(t), x_2(t)]$ now depicts the temporal "comparative advantages" enjoyed in location two in reference to a "numéraire" location, in this instance, location one. We will discuss here the first iterate; the behavior of successive iterates is presented in Section C. Denoting, for simplicity, $x(t) = x_1(t)$ the above dynamics can be represented as

$$x(t + 1) = 1/1 + AF[x(t)], \tag{II.A.2.5}$$

where $F[x(t)] = F_0[x(t), (1 - x(t))]$ and $0 < x(t) < 1$. Thus the next time period's abundance $x(t + 1)$ depends on the current size $x(t)$ only, and the discrete map (see Figure 1) can now be introduced. This map presents the graph of $x(t + 1)$ as a function of $x(t)$. The intersection of this graph with the 45° (diagonal) line, gives the equilibrium state(s) x^*, where $x(t + 1) = x(t)$ for all t's.

b. THE SLOPE AND STABILITY PROPERTIES OF EQUILIBRIA

The slope of the function $x(t + 1)$ on the discrete map, giving the next time period change in the relative population abundance in region one as a function of its current relative size, is presented by

$$s(t + 1, t) = \frac{\partial x(t + 1)}{\partial x(t)} = -Ax^2(t + 1)\frac{\partial F[x(t)]}{\partial x(t)} \tag{II.A.2.6}$$

and it corresponds to the intralocational, intertemporal relative population effect of its current size. We note that $s(t+1, t)$ is a speed of relative stock accumulation is location one. At the equilibrium state x^* the slope $[s^*]$ is

$$s^* = -Ax^{*2}\frac{dF(x^*)}{dx^*}. \tag{II.A.2.7}$$

The behavior of slope s^* defines either the stability or instability of equilibria x^*, and the states in which the iterative process undergoes phase transitions (bifurcations). In the vicinity of the equilibrium state, x^*, the function $x(t+1)$ has the Taylor expansion

$$x(t+1) = x^* + \left.\frac{\partial x(t+1)}{\partial x(t)}\right|_{x^*}[x(t) - x^*] + \frac{1}{2}\left.\frac{\partial^2 x(t+1)}{\partial x(t)^2}\right|_{x^*}[x(t) - x^*]^2 + \cdots. \tag{II.A.2.8}$$

The linear approximation of this expansion is a linear difference equation

$$x(t+1) = x^* + s^*[x(t) - x^*], \tag{II.A.2.9}$$

with a solution given by

$$x(t) = x^* + (s^*)^t[x(0) - x^*]. \tag{II.A.2.10}$$

This implies that equilibrium x^* is (asymptotically) stable ($\lim_{t\to\infty} x(t) = x^*$) if and only if

$$|s^*| < 1. \tag{II.A.2.11}$$

Thus, the expression (II.A.2.11) gives the following description of the domain of stability of equilibria x^*

$$-1 < -Ax^{*2}\frac{dF(x^*)}{dx^*} < 1. \tag{II.A.2.12}$$

Moreover, the equilibria $x^* = x^*(\pm 1)$ with the corresponding slopes $s^* = \pm 1$ are the bifurcation states, i.e., thresholds for phase transitions.

One should note that the derivative $dF(x^*)/dx^*$ is positive at $x^*(-1)$ and negative at $x^*(+1)$

$$\left.\frac{dF(x^*)}{dx^*}\right|_{x^*(-1)} > 0, \qquad \left.\frac{dF(x^*)}{dx^*}\right|_{x^*(+1)} < 0. \tag{II.A.2.13}$$

From the specifications (II.A.2.4) one sees that

$$\frac{\partial x_1(t+1)}{\partial F_0} = -x_1(t+1)^2 A < 0,$$

and at equilibrium (x_1^*, x_2^*)

$$\left.\frac{\partial x_1(t+1)}{\partial F_0}\right|_* = -Ax_1^{*2} < 0. \tag{II.A.2.14}$$

Condition (II.A.2.4) can also be written as

$$\frac{1 - x_1(t+1)}{x_1(t+1)} = \frac{x_2(t+1)}{x_1(t+1)} = AF_0[x_1(t), x_2(t)], \quad \text{(II.A.2.15)}$$

and at the equilibrium state (x_1^*, x_2^*)

$$\frac{x_2^*}{x_1^*} = AF_0^*, \quad \text{(II.A.2.16)}$$

where $F_0^* = F_0(x_1^*, x_2^*)$. The partial slopes, designating interlocational, intertemporal effects, are given by

$$s_{ij}(t+1, t) = \frac{\partial x_i(t+1)}{\partial x_j(t)}; \qquad i, j = 1, 2, \quad \text{(II.A.2.17)}$$

so that in the two-location case

$$s_{11}(t+1, t) = -x_1(t+1)^2 A \frac{\partial F_0}{\partial x_1(t)} = -s_{21}(t+1, t),$$
$$s_{12}(t+1, t) = -x_1(t+1)^2 A \frac{\partial F_0}{\partial x_2(t)} = -s_{22}(t+1, t). \quad \text{(II.A.2.18)}$$

These slopes are the entries of the Jacobi matrix

$$J(t+1, t) = \begin{bmatrix} s_{11}(t+1, t) & s_{12}(t+1, t) \\ s_{21}(t+1, t) & s_{22}(t+1, t) \end{bmatrix}. \quad \text{(II.A.2.19)}$$

Condition $x_1(t+1) + x_2(t+1) = 1$ implies that the Jacobian is zero

$$\det J(t+1, t) = 0. \quad \text{(II.A.2.20)}$$

The trace is

$$\operatorname{Tr} J = s_{11}(t+1, t) + s_{22}(t+1, t)$$
$$= -x_1(t+1)^2 A \left[\frac{\partial F_0}{\partial x_1(t)} - \frac{\partial F_0}{\partial x_2(t)} \right], \quad \text{(II.A.2.21)}$$

and the characteristic equation for the eigenvalues of matrix J is

$$\lambda^2 - \lambda \operatorname{Tr} J + \det J = 0,$$

or

$$\lambda^2 + \lambda x_1(t+1) A \left[\frac{\partial F_0}{\partial x_1(t)} - \frac{\partial F_0}{\partial x_2(t)} \right] = 0, \quad \text{(II.A.2.22)}$$

with the nonzero eigenvalue equal to the slope $s(t+1, t)$

$$\lambda = s_{11}(t+1, t) + s_{22}(t+1, t) = \operatorname{Tr} J$$
$$= -x(t+1)^2 A \frac{dF}{dx(t)} = s(t+1, t). \quad \text{(II.A.2.23)}$$

At equilibrium the Jacobi matrix takes the form

$$J^* = \begin{bmatrix} s_{11}^* & s_{12}^* \\ s_{21}^* & s_{22}^* \end{bmatrix}, \tag{II.A.2.24}$$

where

$$s_{11}^* = -x_1^{*2} \frac{\partial F_0}{\partial x_1}\bigg|_{x^*} = -s_{21}^*,$$
$$s_{12}^* = -x_1^{*2} \frac{\partial F_0}{\partial x_2}\bigg|_{x^*} = -s_{22}^*. \tag{II.A.2.25}$$

The Jacobi matrix, then, has a zero ($\lambda_1 = 0$) and a nonzero ($\lambda_2 \neq 0$) eigenvalue, the latter given by

$$\lambda_2 = s_{11}^* + s_{22}^* = s^*, \tag{II.A.2.26}$$

which is the slope of the discrete map at the equilibrium x^*. Thus, the equilibrium state x^* is stable if the nonzero eigenvalue of the Jacobi matrix at equilibrium satisfies the condition $|\lambda_2| < 1$.

c. Environmental Fluctuations and the Bifurcation Parameter A

Specification (II.A.2.5) implies also that

$$\frac{1 - x(t+1)}{x(t+1)} = AF[x(t)]. \tag{II.A.2.27}$$

At the equilibrium points x^* one has

$$\frac{1 - x^*}{x^*} = AF^*, \tag{II.A.2.28}$$

where $F^* = F(x^*)$. Condition (II.A.2.28) directly indicates that there is a correspondence between any equilibrium state x^* and the parameter A

$$A = \Phi(x^*) = \frac{1 - x^*}{x^* F^*}, \tag{II.A.2.29}$$

linking any equilibrium relative population size to a specific value of (the slow changing environmental fluctuation parameter) A, whereas for any value of the parameter A the equilibrium may not be unique. Conditions (II.A.2.7) and (II.A.2.29) imply that

$$s^* = -x^*(1 - x^*)\frac{d}{dx} \ln F(x^*), \tag{II.A.2.30}$$

which means that the slope s^* can be expressed directly as a function of the equilibria x^*. Condition (II.A.2.30) is central in the analysis that follows.

Taking the derivative of the $\Phi(x^*)$ function from (II.A.2.29) with respect to x^* one obtains

$$\frac{d\Phi(x^*)}{dx^*} = -\frac{1 - s^*}{x^{*2} F^*} = \frac{A(s^* - 1)}{x^*(1 - x^*)}. \tag{II.A.2.31}$$

At the bifurcation states $x^*(\pm 1)$ (with corresponding slopes $s^* = \pm 1$) the values of the parameter $A = A(\pm 1)$, due to (II.A.2.29), are

$$A(-1) = \frac{1 - x^*(-1)}{x^*(-1)F[x^*(-1)]},$$
$$A(+1) = \frac{1 - x^*(+1)}{x^*(+1)F[x^*(+1)]}. \tag{II.A.2.32}$$

Thus, bifurcations take place when A crosses the thresholds $A(-1)$ and $A(+1)$, given that x^* during these crossings falls within the admissible domain $0 < x^* < 1$. Note that

$$A(-1) \to s^* = -1,$$
$$A(+1) \to s^* = +1, \tag{II.A.2.33}$$

and moreover

$$s^* = +1 \quad \to \quad \left.\frac{d\Phi(x^*)}{dx^*}\right|_{x^*(+1)} = 0,$$
$$s^* = -1 \quad \to \quad \left.\frac{d\Phi(x^*)}{dx}\right|_{x^*(-1)} = \frac{-2A(-1)}{x^*(-1)[1 - x^*(-1)]} < 0. \tag{II.A.2.34}$$

In the first case changes in the parameter A have a very severe effect upon the equilibrium population distribution, as the community of (two) regions very near the equilibrium state is extremely sensitive to changes in A. Put differently, in this case it takes a very small fluctuation in the environment (A) to have a significant impact upon the relative stock abundance. Whereas, the second case depicts relative distributions of the stock between the two locations where the fluctuations of A may not have a severe affect.

d. Successive Iterates

Now, the behavior of successive iterates of our general discrete map is presented. The successive iterates $x(t+1), x(t+2), \ldots, x(t+n)$ can be presented with the help of the ith iterates $F^{(i)}$ of the function F (F mapped onto itself i times) in the form

$$x(t+1) = 1/1 + AF^{(1)}[x(t)] \quad \text{with } F^{(1)} = F[x(t)],$$
$$x(t+2) = 1/1 + AF^{(2)}[x(t)] \quad \text{with } F^{(2)} = F[x(t+1)]$$
$$= F\{1/1 + AF[x(t)]\},$$
$$\vdots \qquad \vdots \qquad \vdots$$
$$x(t+n) = 1/1 + AF^{(n)}[x(t)] \quad \text{with a suitable } F^{(n)}[x(t)] = F[x(t+n-1)]. \tag{II.A.2.35}$$

The slope $s(t+1, t) = dx(t+n)/dx(t)$ of the nth iterate $x(t+n)$, $x(t)$ discrete map is given by the chain rule

$$s(t+1,t) = \frac{dx(t+n)}{dx(t)} = \frac{dx(t+n)}{dx(t+n-1)} \cdot \frac{dx(t+n-1)}{dx(t+n-2)} \cdots \frac{dx(t+1)}{dx(t)}$$
$$= s(t+n, t+n-1)s(t+n-1, t+n-2)\dots s(t+1, t). \tag{II.A.2.36}$$

Each $s(t+n-k, t+n-k-1)$ is a function of $x(t+n-k-1)$. Consider now the n-period cycle of the relative dynamics $x(t+1)$, i.e., the sequence of states

$$x^*(0), x^*(1), \dots, x^*(n-1), \tag{II.A.2.37}$$

such that for each integer k

$$x^*(0) = x(t) = x(t+n) = x(t+2n) = \cdots = x(t+kn),$$
$$x^*(1) = x(t+1) = x(t+n+1) = x(t+2n+1) = \cdots = x(t+kn+1),$$
$$\vdots$$
$$x^*(n-1) = x(t+n-1) = x(t+2n-1) = x(t+3n-1) = \cdots$$
$$= x(t+kn+n-1). \tag{II.A.2.38}$$

Each state $x^*(j)$, $j = 0, 1, 2, \dots, n-1$, is a fixed point of the nth iterate $x(t+n)$

$$x^*(j) = 1/1 + F^{(n)}[x^*(j)], \qquad j = 0, 1, 2, \dots, n-1. \tag{II.A.2.39}$$

At each state $x^*(j)$ of an n-period cycle, the discrete map for $x(t+n)$ intersects the 45° straight line. The slope

$$s_n^*[x^*(j)] = s(t+n, t)|_{x^*(j)} \tag{II.A.2.40}$$

in each of these intersections is

$$s_n^*[x^*(j)] = s(t+n, t+n-1)|_{x^*(j)} s(t+n-1, t+n-2)|_{x^*(j)} \cdots s(t+1, t)|_{x^*(j)}$$
$$= s^*[x^*(j-1)]s^*[x^*(j-2)]\dots s^*[x^*(j)], \tag{II.A.2.41}$$

independently of j. Obviously, the condition of stability for an n-period cycle $[x^*(0), x^*(1), \dots, x^*(n-1)]$ is

$$|s_n^*| = |s^*[x^*(0)]s^*[x^*(1)]\dots s^*[x^*(n-1)]| < 1. \tag{II.A.2.42}$$

This concludes our analysis for the iterative behavior of our general map.

3. *Competitive Exclusion Equilibria*

The study at the neighborhood of the improper equilibria, ($x_1^* = 0$, $x_2^* = 1$), ($x_1^* = 1$, $x_2^* = 0$), is of interest since it gives some insight into the special behavior of the universal map. The behavior of the function $x(t+1)$ near the end points of the interval (0, 1) is defined by the values of the one-sided limits

$$x_0 = \lim_{x(t)\to 0+} x(t+1); \qquad x_1 = \lim_{x(t)\to 1-} x(t+1),$$

$$s_0 = \lim_{x(t)\to 0+} s[t+1, t]; \qquad s_1 = \lim_{x(t)\to 1-} s[t+1, t].$$

Limits of $x(t+1)$ define the conditions under which x^* approaches either zero or one, whereas the limits of s define their stability/instability properties. There are three values which the $\lim F[x(t)]$ can attain

$$\lim_{x(t)\to 0+} F[x(t)] = \begin{cases} +\infty \\ c_1 & (0 < c_1 < +\infty), \\ 0 \end{cases}$$

therefore

$$x_0 = \lim_{x(t)\to 0+} x(t+1) = \begin{cases} 0, \\ 1/1 + Ac_1, \\ 1. \end{cases} \tag{II.A.3.1}$$

Similarly

$$\lim_{x(t)\to 1-} F[x(t)] = \begin{cases} +\infty \\ c_2 & (0 < c_2 < +\infty), \\ 0 \end{cases}$$

and thus

$$x_1 = \lim_{x(t)\to 1-} x(t+1) = \begin{cases} 0, \\ 1/1 + Ac_2, \\ 1. \end{cases} \tag{II.A.3.2}$$

The competitive exclusion equilibrium $x^* = 0$ (or $x^* = 1$) is stable if $|s_0| < 1$ (or $|s_1| < 1$.) Thus, any equilibrium point arbitrarily close to competitive exclusion could be either a repeller or an attractor. Put differently, one can find $\hat{A}$'s such that competitive exclusion is approached at any arbitrarily defined neighborhood of the boundaries zero or one. This, as will be seen later, has some implications for the results to be obtained by numerical simulations, and their interpretation.

4. *Classification of Fundamental Relative Spatial Dynamics*

The interdependencies (II.A.2.29–31) among the comparative advantages function $F(x^*)$, the slope s^*, the parameter $A = \Phi(x^*)$, and the equilibrium values of the iterative process, x^*, provide the basis for classifying the various families of the functions F of the universal one-(homogeneous) stock, two-(heterogeneous) region relative dynamics.

From an analytical (computational) standpoint the choice of F can be such that the most general results can be derived, given some specific dynamic phenomena which one wishes to investigate. The focus of this analysis is the study of conditions under which the phenomenon of turbulence in relative spatial dynamics can occur.

Classification can be carried out in terms of either specifying a comparative advantage function F, then computing s^* through (II.A.2.30) and finally examining the behavior of x^* through (II.A.2.29). Or, it can be obtained by setting an analytical form for the s^* function, and then deriving the comparative advantage F and Φ^* functions. The formulas are as follows

$$F(x^*) = \exp\left[-\int \frac{s^*\,dx^*}{x^*(1-x^*)}\right], \tag{II.A.4.1}$$

$$\Phi(x^*) = \frac{1-x^*}{x^*F(x^*)} = A. \tag{II.A.4.2}$$

We start with a polynomial-type s^* function. This family of functions includes two types which will be examined in detail: a log-linear comparative advantages producing function, and an exponential type. Consider the polynomial function

$$s^* = \sum_{k=0}^{K} a_k x^{*k}, \tag{II.A.4.3}$$

with an integrand of its $F(x^*)$ function specified as

$$\begin{aligned}\frac{-s^*}{x^*(1-x^*)} &= -\sum_{k=0}^{K} \frac{a_k x^{*k}}{x^*(1-x^*)}\\ &= \sum_{k=2}^{K} (a_K + a_{K-1} + \cdots + a_{K-k+2})x^{*K-k}\\ &\quad -\frac{a_0}{x^*} - \frac{a_K + \cdots + a_1 + a_0}{1-x^*},\end{aligned}$$

and therefore

$$\begin{aligned}&-\int \frac{s^*\,dx^*}{x^*(1-x^*)}\\ &= \sum_{k=2}^{K} \frac{(a_K + \cdots + a_{K-k+2})x^{*K-k+1}}{K-k+1} - a_0 \ln x^* + (a_K + \cdots + a_0)\ln(1-x^*).\end{aligned} \tag{II.A.4.4}$$

The above gives, from (II.A.4.1, 2),

$$F(x^*) = \left[\exp\left(\sum_{k=2}^{K} \frac{a_K + \cdots + a_{K-k+2}}{K-k+1} x^{*K-k+1}\right)\right] x^{*-a_0}(1-x^*)^{a_K+\cdots+a_0}, \tag{II.A.4.5}$$

$$A = \Phi(x^*) = \frac{x^{*-(1+a_0)}(1-x^*)^{1-(a_K+\cdots+a_0)}}{\exp\left(x^* \sum_{k=2}^{K} \frac{a_K + \cdots + a_{K-k+2}}{K-k+1} x^{*K-k}\right)}, \tag{II.A.4.6}$$

and, finally, deriving the comparative advantage function giving rise to this s^* function, one has

$$F[x(t)] = x(t)^{-a_0}[1 - x(t)]^{a_K + \cdots + a_0} \exp\left[x(t) \sum_{k=2}^{K} \frac{a_K + \cdots + a_{K-k+2}}{K - k + 1} x(t)^{K-k}\right]. \tag{II.A.4.7}$$

We will return to this statement when discussing the log-linear specifications of F in subsection 2 of Section B below.

So far, simple exponential comparative advantages producing functions of the type $F[x(t)] = \exp\{P[x(t)]\}$, where $P(x)$ is a polynomial of x, were considered. A more general case of the type

$$F[x(t)] = \exp\left\{\sum_{k=1}^{K} C_k x(t)^{a_k}[1 - x(t)]^{b_k}\right\} \tag{II.A.4.8}$$

can also be examined. Its corresponding parameter function $A = \Phi(x^*)$ calculated from (II.A.2.29) is

$$A = \Phi(x^*) = \frac{1 - x^*}{x^* F(x^*)}$$
$$= \frac{1 - x^*}{x^*} \exp\left[-\sum_{k=1}^{K} C_k x^{*a_k}(1 - x^*)^{b_k}\right]. \tag{II.A.4.9}$$

From (II.A.2.30) its corresponding slope function s^* is given by

$$s^* = -x^*(1 - x^*)\frac{d}{dx^*} \ln F(x^*)$$
$$= \sum_{k=1}^{K} C_k x^{*a_k}(1 - x^*)^{b_k}[(a_k + b_k)x^* - a_k], \tag{II.A.4.10}$$

which is a generalized log-linear specification with linear multipliers. It is possible to extend this classification even further, but this extension can only be carried out if it is warranted by substantially important cases in particular real stock dynamics.

B. Log-Linear Comparative Advantages Producing Functions: $F = x(t)^a[1 - x(t)]^b$.

1. *Interpretation and Discussion of the Log-Linear Model*

a. Introduction

In this section we specify, interpret, and present in more detail the log-linear specifications of the function F, subsection 1; intervals of stability for its *first iterate* equilibria are presented in subsection 2; the analytical properties are discussed in subsection 3; with a geometric presentation given in subsection 4.

b. SPECIFICATIONS AND DEFINITIONS

We now specify the temporal and locational advantages generating production functions F, as two separate log-linear functions

$$F_1[x_1(t), x_2(t)] = x_1(t)^{\alpha_1} x_2(t)^{\alpha_2}, \tag{II.B.1.1}$$

$$F_2[x_1(t), x_2(t)] = x_1(t)^{\beta_1} x_2(t)^{\beta_2}, \tag{II.B.1.2}$$

$$0 < x_1(t), x_2(t) < 1,$$

$$-\infty \leq \alpha_1, \alpha_2, \beta_1, \beta_2 \leq +\infty.$$

These broad specifications of socio-spatial dynamics draw from references to both classical and general location theory, and from the economic theory of production and comparative advantages specifically. The current locational advantages enjoyed by any social stock found in both regions depend on the current relative stock (population) size at these two locations, $x_1(t), x_2(t)$. They also depend upon intrinsic characteristics of these regions depicted by the parameters (exponents) of the above general socio-spatial model. The vector of state variables, $x_1(t)$, $x_2(t)$, represent the current configuration of the relative stock (population) distribution or, in other words, the current relative socio-spatial configuration.

Parameters depict composite socioeconomic–geographic factors. These parameters may depend, among other factors, on economic forces associated with the relative availability of natural resources in these regions, agglomeration (scale) effects in each location, etc.; as well as on their geographic features like topography, climate, distance relative to one another, etc. They may also depend on a host of other social (broadly referred to as cultural) factors involved in generating temporal locational advantages for attracting the stock (population) in question.

The exponents of these F functions represent elasticities of locational advantages generated in each region with respect to the current relative regional stock sizes. The magnitude of their two sums is of interest

$$\alpha = \alpha_1 + \alpha_2, \qquad \beta = \beta_1 + \beta_2, \tag{II.B.1.3}$$

since, when $\alpha = 1$ (or $\beta = 1$), there are constant returns-to-scale in the production of temporal locational advantages in location one (α) or location two (β), given a particular configuration of the relative population distribution. The difference

$$\delta = \beta - \alpha \tag{II.B.1.4}$$

identifies the returns-to-scale differential between the two regions, and much of the qualitative properties of the discrete general socio-spatial dynamics depend on that difference. These parameters will be referred to as "structural" parameters.

According to this model formulation, stock $x(t)$, and its antiregional equivalent $x^-(t) = 1 - x(t)$, adjusts dynamically in such a way that each succeeding generation's stock at a given region depends on the comparative advantages enjoyed in that region by the preceding generation. Thus, the general one-time-period delay mechanism of this formalism depicts the temporal dependence in the central state variable, which in this case is the relative size of the homogeneous stock (population). The above model can now be written as

$$x_1(t+1) = \frac{A_1 x_1(t)^{\alpha_1} x_2(t)^{\alpha_2}}{A_1 x_1(t)^{\alpha_1} x_2(t)^{\alpha_2} + A_2 x_1(t)^{\beta_1} x_2(t)^{\beta_2}}, \qquad t = 1, 2, \ldots, T,$$

$$x_2(t+1) = \frac{A_2 x_1(t)^{\beta_1} x_2(t)^{\beta_2}}{A_1 x_1(t)^{\alpha_1} x_2(t)^{\alpha_2} + A_2 x_1(t)^{\beta_1} x_2(t)^{\beta_2}}, \qquad 0 < x_1(0), \quad x_2(0) < 1, \tag{II.B.1.5}$$

where the parameter vector $\bar{A} = (A_1, A_2)$ is associated with environmental fluctuations, technological innovations, etc. It follows directly that

$$x_1(t+1) + x_2(t+1) = 1, \qquad t = 1, 2, \ldots, T.$$

Formulating the above problem in terms of comparative advantages, and dropping the subscript, one obtains

$$x(t+1) = 1/1 + Ax(t)^a[1 - x(t)]^b, \qquad 0 < x(t+1) < 1, \quad \text{(II.B.1.6)}$$

where the differences, $(a = \beta_1 - \alpha_1)$ and $(b = \beta_2 - \alpha_2)$, represent differentials in each of the two regions' locational advantages elasticities with respect to the regional stock size. Note that $\delta = a + b$, so that the returns-to-scale in temporal comparative advantages is equal to the regions' returns-to-scale in the locational advantages differential. In this formulation, there is one (relative) environmental fluctuation parameter, A, given by $A = A_2/A_1$; it depicts the end-effect of various environmental perturbations in the two regions.

Going back to conditions (II.B.1.1, 2) it is noted that in the sign of the four exponents in the F_1 and F_2 functions, all possible locational associations are embedded for the single stock considered. These associations are more complex than the six simple ecological associations (competition, cooperation, predation, isolation, amensalism, commensalism) found in the field of mathematical ecology according to the Volterra–Lotka formalism (May, 1974). In the Volterra–Lotka model the six basic associations depend upon the combination of signs among the inter- and intraspecies interaction coefficients. Here, the various types of associations existing between two locations depend on both the sign and size of the exponent coefficients in the iterative map. Isolation, amensalism, and commensalism are not feasible under the universal discrete relative dynamics map; whereas the pure form of the remaining three basic ecological associations vanishes in the general case. Instead, various combinations of the competition, cooperation, and predation modes are obtained, dependent upon the map's specific position in its parameter space. Under discrete dynamics a new classification emerges for stock-location temporal associations. A classification of such behavior is supplied in subsequent sections of this Part.

c. The Three Speeds of the Model

The one-period *adjustment* process identifies relatively fast dynamics either toward or away from the equilibrium point. Changes in the parameters, a and b, represent relatively *very slow* dynamics. It is the random and *rather slow* changes in the environmental fluctuations, A, that are responsible for the interesting qualitative properties of the system's dynamic behavior. Due to its role, parameter A will be referred to as the "bifurcation parameter."

This model can be conceived, too, as a zero-sum game between two activities (or regions) competing for a fixed resource (homogeneous stock), like population. A potentially interesting example of this fixed-point process, in economic growth theory, might be the relatively fast dynamical behavior of output shares going into current consumption or production, the latter in the form of savings during the current time period and investment during the next time period. In this example, the "two-region" setting has been substituted by a "two-sector" specification. Economic evolution then can be viewed as cyclical, logistic, or chaotic adjustment of these two stocks over time, as relatively slow movements occur in the structural and bifurcation parameters.

2. *Intervals of Stability of Equilibria*

The most important property of the log-linear model (II.B.1.6) is that its comparative advantage function $F[x(t)] = x(t)^a[1 - x(t)]^b$ belongs to the slope polynomial type (II.A.4.3) with $a_0 = -a$, $a_1 = a + b$, and $a_2 = a_3 = \cdots = a_K = 0$. This means that the log-linear model has a *linear*, in x^*, s^* function

$$s^* = (a + b)x^* - a. \qquad \text{(II.B.2.1)}$$

The domain of stability of the equilibrium x^* in the space of the parameters, a and b, is defined by the inequality

$$|s^*| = |(a + b)x^* - a| < 1 \qquad \text{(II.B.2.2)}$$

or

$$-1 < (a + b)x^* - a < 1. \qquad \text{(II.B.2.3)}$$

Geometrically, this inequality determines an interval, whose end points $x^*(\pm 1)$ correspond to the slopes $s^* = \pm 1$. The values of $x^*(\pm 1)$ at these bifurcation points are

$$s^* = 1 \quad \rightarrow \quad (a + b)x^*(+1) - a = 1 \quad \rightarrow \quad x^*(+1) = \frac{a + 1}{a + b}; \qquad \text{(II.B.2.4)}$$

$$s^* = -1 \quad \rightarrow \quad (a + b)x^*(-1) - a = -1 \quad \rightarrow \quad x^*(-1) = \frac{a - 1}{a + b}; \qquad \text{(II.B.2.5)}$$

The domain of stability of equilibria is defined by the intersection of the interval (0, 1) with the interval $[x^*(-1), x^*(+1)]$ or $[x^*(+1), x^*(-1)]$ (see

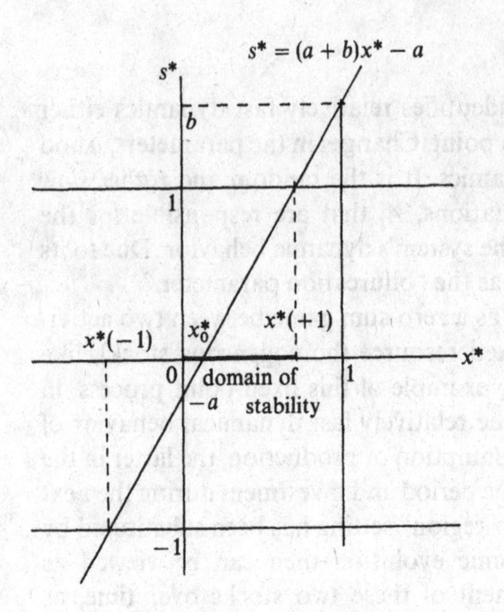

FIGURE 2. The domain of stability of equilibria as the intersection of the interval (0, 1) with the interval $[x^*(-1), x^*(+1)]$.

Figure 2). Further, since $s^*(0) = -a$, $s^*(1) = b$, then

$$\begin{aligned} -a < s^* < b \quad &\text{for} \quad a + b > 0 \\ b < s^* < -a \quad &\text{for} \quad a + b < 0. \end{aligned} \tag{II.B.2.6}$$

From the above two requirements it follows immediately that in the case $|a|, |b| < 1$ the slope at equilibrium $|s^*|$ is less than one, and the equilibrium is stable (and unique). Thus, within this domain x^* is always a fixed-point attractor so that this region of the parameter space is an area of superstability. Note that both the constant returns-to-scale case *for the comparative advantages*, $a + b = 1$, and the case where the differential in the two regional stocks elasticities with respect to *locational* advantages in the two regions is equal to one, $\delta = 1$, fall under the $|a|, |b| < 1$ dynamic.

3. *Analytical Properties of the Log-Linear Model*

a. The Parameter Function $A = \Phi(x^*)$ and Its Properties

Expression (II.A.4.2) gives the equation for computing the values of equilibria x^*

$$A = \Phi(x^*) = (x^*)^{-(a+1)}(1 - x^*)^{1-b}, \tag{II.B.3.1}$$

showing the one-to-one correspondence from x^* to A, and the possible one-to-many correspondence from A to x^*. This can be expressed as a mapping

$$\Phi = (x^* \rightarrow A; a, b \in R^2).$$

Condition (II.B.3.1) states that, given any (stable or unstable) equilibrium arbitrarily drawn from the admissible open interval (0, 1) and an arbitrary set of a, b coefficients, there is an environmental parameter value A which would make this spatial configuration eventually possible. In other words, equation (II.B.3.1) means that the bifurcation parameter, A, can be obtained as a function of the equilibria x^*: $A = \Phi(x^*)$. Moreover, the graph of this function enables one to construct its inverse function $x^* = \Phi^{-1}(A)$, and to define the qualitative properties and number of equilibrium states for the log-linear model.

The following properties will be useful for the analysis to follow:

let $0 < x < 1$ then

$$\lim_{x \to 0+} x^{c_1}(1-x)^{c_2} = \begin{cases} 0, & c_1 > 0, \\ 1, & c_1 = 0, \\ +\infty, & c_1 < 0, \end{cases} \tag{II.B.3.2}$$

and

$$\lim_{x \to 1-} x^{c_1}(1-x)^{c_2} = \begin{cases} 0, & c_2 > 0, \\ 1, & c_2 = 0, \\ +\infty, & c_2 < 0, \end{cases} \tag{II.B.3.3}$$

Indeed, let $f(x) = x^{c_1}(1-x)^{c_2}$, then $z = \ln x = c_1 \ln f + c_2 \ln(1-x)$. We observe that

$$\lim_{x \to 0+} z = \begin{cases} -\infty, & c_1 > 0, \\ 0, & c_1 = 0, \\ +\infty, & c_1 < 0, \end{cases}$$

and

$$\lim_{x \to 1-} z = \begin{cases} -\infty, & c_2 > 0, \\ 0, & c_2 = 0, \\ +\infty, & c_2 < 0, \end{cases}$$

which immediately imply (II.B.3.2, 3).

The explicit form (II.B.3.1) of function $\Phi(x^*)$ allows one to draw its graph, as the properties (II.B.3.2, 3) give the description of functions $\Phi(x^*)$ at the end points 0, 1

$$\lim_{x^* \to 0+} \Phi(x^*) = \begin{cases} 0, & a < -1, \\ 1, & a = -1, \\ +\infty, & a > -1, \end{cases} \tag{II.B.3.4}$$

and

$$\lim_{x^* \to 1-} \Phi(x^*) = \begin{cases} 0, & b < 1, \\ 1, & b = 1, \\ +\infty, & b > 1. \end{cases} \tag{II.B.3.5}$$

Consequently, we have nine different types of graphs, each corresponding to a behavior for the function $\Phi(x^*)$ (see Figure 3).

Further, from (II.A.2.31) one sees that the function $\Phi(x^*)$ has a singular (maximum/minimum/inflection) point at $x^*(+1)$ which is obtained from the

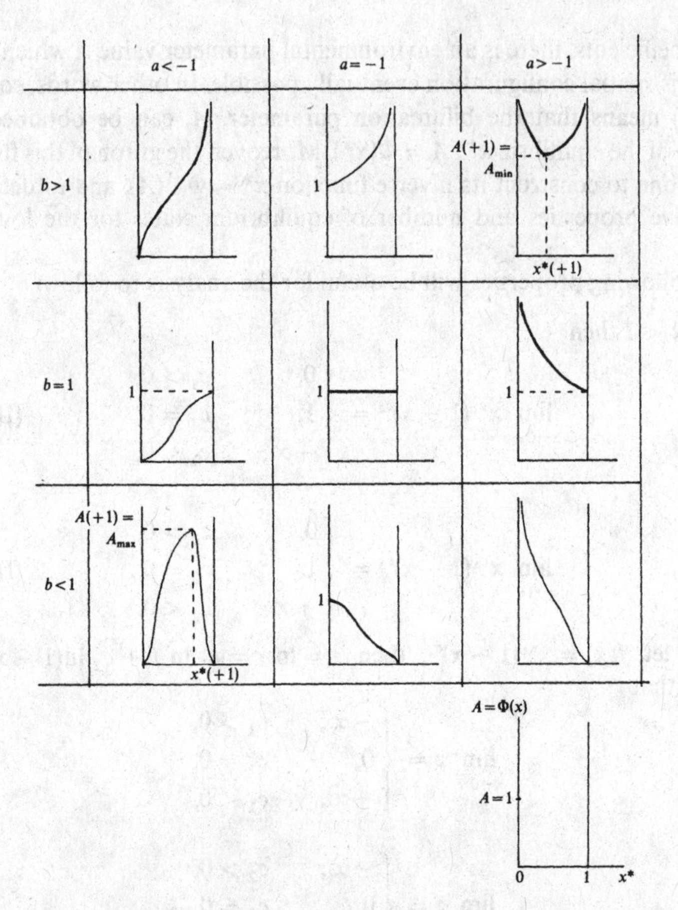

FIGURE 3. The graph of the function $A = \Phi(x^*)$.

first and second derivatives

$$\frac{d\Phi(x^*)}{dx^*} = -A\left(\frac{a+1}{x^*} + \frac{1-b}{1-x^*}\right) = \frac{A(s^*-1)}{x^*(1-x^*)},$$
$$\frac{d^2\Phi(x^*)}{dx^{*2}} = A\left[\left(\frac{a+1}{x^*} + \frac{1-b}{1-x^*}\right)^2 + \left(\frac{a+1}{x^{*2}} + \frac{b-1}{(1-x^*)^2}\right)\right]. \qquad \text{(II.B.3.6)}$$

Values of the parameter A, where bifurcation occurs in the behavior of x^*, are

$$A(+1) = \Phi[x^*(+1)] = [x^*(+1)]^{-(a+1)}[1-x^*(+1)]^{1-b}$$
$$= \left[\frac{a+1}{a+b}\right]^{-(a+1)}\left[\frac{b-1}{a+b}\right]^{1-b}, \qquad \text{(II.B.3.7)}$$
$$A(-1) = \Phi[x^*(-1)] = [x^*(-1)]^{-(a+1)}[1-x^*(-1)]^{1-b}$$
$$= \left[\frac{a-1}{a+b}\right]^{-(a+1)}\left[\frac{b+1}{a+b}\right]^{1-b}. \qquad \text{(II.B.3.8)}$$

The requirement that $0 < x^*(+1) < 1$ necessitates that the following conditions must hold when bifurcations occur

$$a > -1, \quad b > 1, \tag{II.B.3.9.1}$$

or

$$a < -1, \quad b < 1. \tag{II.B.3.9.2}$$

Note that

$$\left.\frac{d^2\Phi(x^*)}{dx^{*2}}\right|_{x^*(+1)} = A(+1)\frac{(a+b)^3}{(a+1)(b-1)}, \tag{II.B.3.10}$$

and when $a > -1, b > 1$, this second derivative is positive, i.e., $x^*(+1)$ results in a minimum for $\Phi(x^*)$

$$A_{\min} = \frac{(a+b)^{a+b}}{(a+1)^{a+1}(b-1)^{b-1}} = A(+1). \tag{II.B.3.11}$$

Analogously, when $a < -1, b < 1$, the second derivative is negative, and $x^*(+1)$ gives a maximum for $\Phi(x^*)$

$$A_{\max} = \frac{|a+b|^{a+b}}{|a+1|^{a+1}|b-1|^{b-1}} = A(+1). \tag{II.B.3.12}$$

Obviously,

$$\frac{d\Phi(x^*)}{dx^*} = -A\left[\frac{a+1}{x^*} + \frac{1-b}{1-x^*}\right]$$

is positive in the domains

$$a < -1, \quad b > 1, \tag{II.B.3.13.1}$$

$$a = -1, \quad b > 1, \tag{II.B.3.13.2}$$

$$a < -1, \quad b = 1; \tag{II.B.3.13.3}$$

and negative in the domains

$$a > -1, \quad b < 1, \tag{II.B.3.14.1}$$

$$a = -1, \quad b < 1, \tag{II.B.3.14.2}$$

$$a > -1, \quad b = 1; \tag{II.B.3.14.3}$$

thus providing the regions where monotonic growth or decline occurs in the function $\Phi(x^*)$.

Moreover, the second derivative $d^2\Phi(x^*)/dx^{*2}$ is positive in the domains

$$a > -1, \quad b > 1, \tag{II.B.3.15.1}$$

$$a = -1, \quad b > 1, \tag{II.B.3.15.2}$$

$$a > -1, \quad b = 1; \tag{II.B.3.15.3}$$

i.e., in these domains $\Phi(x^*)$ is concave. On the basis of the above considerations one can construct the graphs of Figure 3 allowing one to geometrically obtain the conditions for existence of equilibria, and also to compute the number of equilibrium states.

b. THE INVERSE FUNCTION $x^* = \Phi^{-1}(A)$

Figure 4 presents a graph of the inverse function $x^* = \Phi^{-1}(A)$ in each of the nine domains (II.B.3.13.1–15.3); it is symmetrical to Figure 3 with respect to the 45° line. It demonstrates the conditions for existence, and reveals the number of equilibrium states and their dynamics following changes in the environmental parameter A. For example, in the domain, $a < -1, b < 1$, there is a maximal value of the parameter A: $A_{max} = A(+1)$, such that for each $A < A(+1)$ there are two equilibrium states $(x^*_{1,1}, x^*_{1,2} = 1 - x^*_{1,1})$ and $(x^*_{2,1}, x^*_{2,2} = 1 - x^*_{2,1})$. The values of the equilibria start to change at $(x^*_{1,1} = 0$ and $x^*_{1,2} = 1)$ and $(x^*_{2,1} = 1, x^*_{2,2} = 0)$. As A changes from zero to $A(+1)$ the equilibria values $x^*_{1,1}, x^*_{2,1}$ move in the opposite direction toward the bifurcation state $x^*(+1) = (a + 1)/(a + b)$, so that they merge with $x^*(+1)$ at $A = A(+1)$. If $A > A(+1)$ then equilibria do not exist.

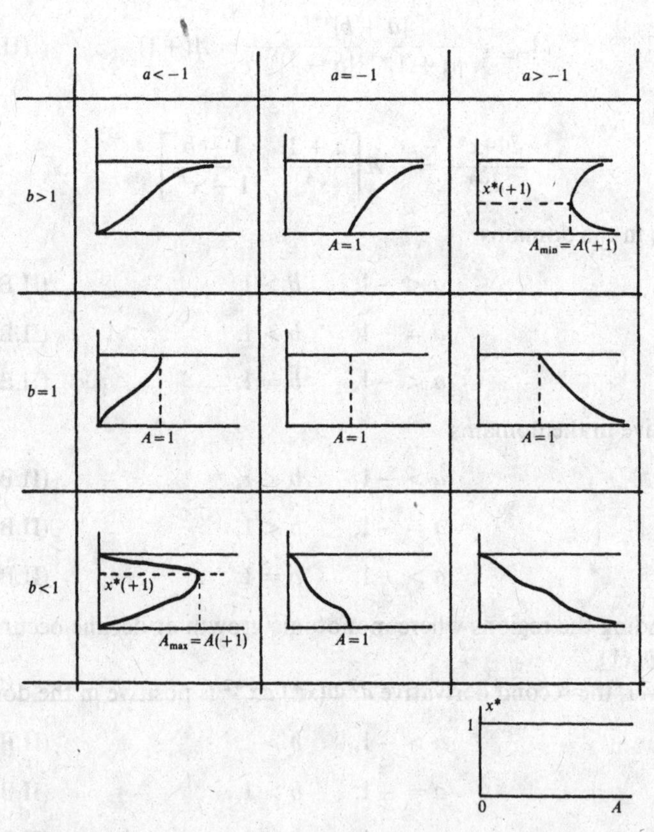

FIGURE 4. Graphical representation of the equilibria x^* of the first iterate, as functions of A.

4. *Geometric Description of the Iterative Process*

a. Behavior at the Limits

Next we consider the behavior of the iterative process in the vicinity of the end points 0 and 1. This enables us to obtain the conditions for existence and stability of competitive exclusion equilibria. It also allows us to classify the complete catalogue of (49) cases of behavior in our special version of the universal relative iterative mapping.

Expressions (II.B.3.2, 3) imply the following behavior of the universal iterative process (II.B.1.6) at the limits 0 and 1

$$\lim_{x(t)\to 0+} x(t+1) = \begin{cases} 1, & a > 0, \\ 1/(1+A), & a = 0, \\ 0, & a < 0, \end{cases}$$
$$\lim_{x(t)\to 1-} x(t+1) = \begin{cases} 1, & b > 0, \\ 1/(1+A), & b = 0, \\ 0, & b < 0. \end{cases} \qquad \text{(II.B.4.1)}$$

The complete behavior at the limits is depicted in Figure 5 for the combined changes in the structural parameters a, b. It is of interest to note that if the environmental fluctuation parameter A obtains appropriate values, it may offset the effect that differentials, in elasticities of locational advantages, have

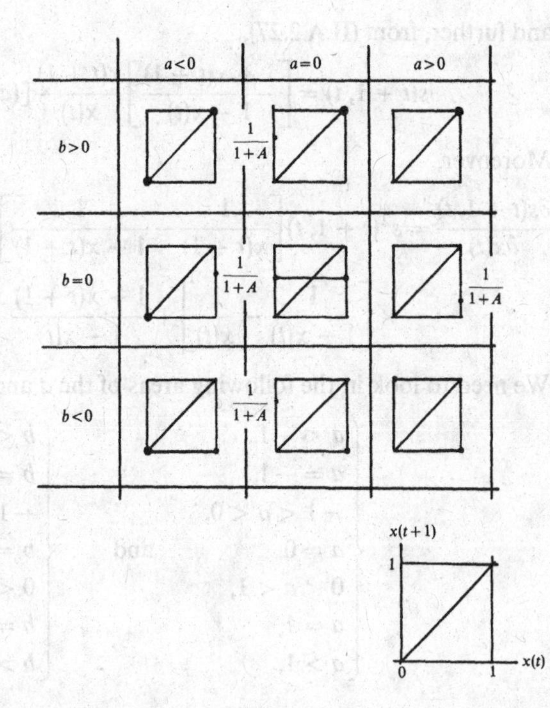

FIGURE 5. Limit behavior. The behavior of the iterative process $x(t+1) = 1/1 + Ax(t)^a[1 - x(t)]^b$ at the end points for nine domains of the (a, b) space; dots represent the limits of $x(t+1)$ as $x(t)$ approaches zero, or one; bigger dots represent the competitive exclusion equilibria.

in the two regions. For example, if $A \gg 0$ then the transition from $a < 0$ to $a > 0$, for $b < 0$, can be quite abrupt as the $[x(t+1), x(t) = 0]$ point might jump from zero to one; whereas the transition may be smooth if simply A is positive but small.

b. The 49 Cases of Behavior

In this subsection, the 49 different types of associations are presented. They identify all possible associations among locations in the case of a unique (homogeneous) stock inhabiting them. These associations are far more complicated than the simple six ecological associations detected in the Volterra–Lotka models of mathematical ecology, which is defined in continuous dynamics.

The slope of $x(t+1)$ in the $[x(t+1), x(t)]$ space is given by, see (II.A.2.6),

$$s(t+1, t) = -Ax(t+1)^2 \frac{\partial F[(t)]}{\partial x(t)}.$$

Obviously, in the log-linear formulation, one has

$$\frac{\partial F[x(t)]}{\partial x(t)} = x(t)^{(a-1)}[1 - x(t)]^{(b-1)}\{a[1 - x(t)] - bx(t)\}, \quad \text{(II.B.4.2)}$$

so that

$$s(t+1, t) = -Ax(t+1)^2 \frac{F[x(t)]}{x(t)[1 - x(t)]}\{a[1 - x(t)] - bx(t)\}, \quad \text{(II.B.4.3)}$$

and further, from (II.A.2.27),

$$s(t+1, t) = \left[\frac{1 + x(t+1)}{1 - x(t)}\right]\frac{x(t+1)}{x(t)}[(a+b)x(t) - a]. \quad \text{(II.B.4.4)}$$

Moreover,

$$\frac{\partial s(t+1, t)}{\partial x(t)} = s^2(t+1, t)\left[\frac{1}{x(t+1)} - \frac{1}{1 - x(t+1)}\right] + s(t+1, t)$$
$$\times \left[\frac{1}{1 - x(t)} - \frac{1}{x(t)}\right] + \frac{1 - x(t+1)}{1 - x(t)}\frac{x(t+1)}{x(t)}(a+b). \quad \text{(II.B.4.5)}$$

We need to look in the following areas of the a and b ranges

$$\begin{cases} a < -1, \\ a = -1, \\ -1 < a < 0, \\ a = 0, \\ 0 < a < 1, \\ a = 1, \\ a > 1, \end{cases} \quad \text{and} \quad \begin{cases} b < -1, \\ b = -1, \\ -1 < b < 0, \\ b = 0, \\ 0 < b < 1, \\ b = 1, \\ b > 1, \end{cases}$$

creating in all 49 cases to be analyzed. From (II.B.4.4) and (II.B.1.6) one obtains

$$s(t+1,t)=\frac{\partial x(t+1)}{\partial x(t)}$$

$$=\frac{A[(a+b)x(t)-a]}{\{x(t)^{(1-a)/2}[1-x(t)]^{(1-b)/2}+Ax(t)^{(a+1)/2}[1-x(t)]^{(b+1)/2}\}^2}, \tag{II.B.4.6}$$

with limits given by

$$\lim_{x\to 0+} s(t+1,t)=\begin{cases}0, & a>1,\\ -A, & a=1,\\ -\infty, & 0<a<1,\\ Ab/(1+A)^2, & a=0,\\ +\infty, & -1<a<0,\\ 1/A, & a=-1,\\ 0, & a<-1,\end{cases} \tag{II.B.4.7}$$

$$\lim_{x\to 1-} s(t+1,t)=\begin{cases}0, & b>1,\\ A, & b=1,\\ +\infty, & 0<b<1,\\ -Aa/(1+A)^2, & b=0,\\ -\infty, & -1<b<0,\\ -1/A, & b=-1,\\ 0, & b<-1.\end{cases} \tag{II.B.4.8}$$

To complete the discrete map one needs to find the maximum/minimum points for $x(t+1)$. Note that $s(t+1,t)$ is zero, i.e., $x(t+1)$ becomes stationary, if and only if the sum of the returns-to-scale differences between the two regions is nonzero ($\delta=a+b\neq 0$). Thus

$$s(t+1,t)=0 \quad\rightarrow\quad 0<\tilde{x}(t)=\frac{a}{a+b}=x_0^*<1. \tag{II.B.4.9}$$

The above demonstrates that $x(t+1)$ attains *only one* singular point x_0^*; and this point exists in the admissible domain (0, 1) if and only if

$$\text{(a)}\quad a>0 \quad\rightarrow\quad a+b>0 \quad\text{and}\quad b>0, \tag{II.B.4.10}$$

$$\text{(b)}\quad a<0 \quad\rightarrow\quad a+b<0 \quad\text{and}\quad b<0, \tag{II.B.4.11}$$

i.e., the differences in each of the two regions' locational advantages elasticities with respect to the regional population (a, b) are both either positive or negative.

The value of $\tilde{x}(t+1)$ of the function $x(t+1)$ in the singular point $\tilde{x}(t) = x_0^* = a/(a+b)$ is

$$\tilde{x}(t+1) = 1/1 + A|a|^a|b|^b|a+b|^{-(a+b)}. \quad \text{(II.B.4.12)}$$

Note that if $a = b$, then $x_0^* = \frac{1}{2}$ and $\tilde{x}(t+1) = (1 + A/4)^{a-1}$. As $A \to 0$, $\tilde{x}(t+1) \to 1$, and as $A \to \infty$, $\tilde{x}(t+1) \to 0$.

From (II.B.4.5) we obtain

$$\left.\frac{\partial s(t+1,t)}{\partial x(t)}\right|_{x_0^*} = \left[\frac{1-\tilde{x}(t+1)}{1-x_0^*}\right]\frac{\tilde{x}(t+1)}{x_0^*}(a+b). \quad \text{(II.B.4.13)}$$

This means that in the domain (II.B.4.10) $x(t+1)$ has a minimum, and in the domain (II.B.4.11) it obtains a maximum. The dynamics of the changes in the number of equilibria, and the dynamics of transitions from stability to instability will be demonstrated next for a special choice of the structural parameters a, b in the log-linear model, $a > 1$, $b > 1$.

c. A Special Case ($a > 1, b > 1$)

The domain, $a > 1$, $b > 1$, is only one of 49 different domains of the possible ranges in the parameters a, b. Each is associated with different types of equilibria with distinct qualitative features. In this case the function $\Phi(x^*)$ has a minimum at point $x^*(+1) = (a+1)/(a+b)$ and where $A_{\min} = A(+1)$ (see Figure 6). Therefore, for $A < A(+1)$ there are no equilibrium states. For

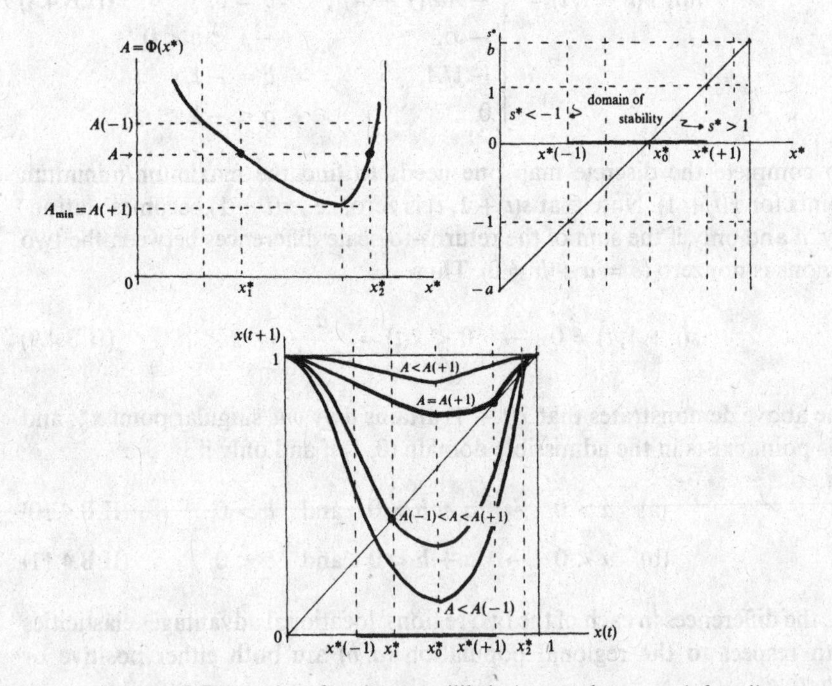

FIGURE 6. Dynamics of various equilibria (cases where $a > 1$, $b > 1$).

$A = A(+1)$ there is a bifurcation state $x^*(+1)$ with the corresponding slope $s^* = +1$. This equilibrium has an interesting property: it is an attractor from the left and a repeller from the right. If $A(+1) < A < A(-1)$ then two equilibrium states appear: one stable x_1^*: $x^*(-1) < x_1^* < x^*(+1)$, and one unstable x_2^*: $x^*(+1) < x_2^*$, which repels the subsequent states of $x(t)$ eventually producing the competitive exclusion state, $x^* = 1$. If A further increases $[A \geq A(-1)]$, then the stable equilibrium x_1^* crosses the bifurcation state $x^*(-1)$ and becomes unstable. This particular region of the (a, b) space affords the possibility for period-doubling cycles, as shown next.

Combining the cases found in conditions (II.B.4.7–11) and the conditions guiding existence and evolution of the various types of equilibria shown in

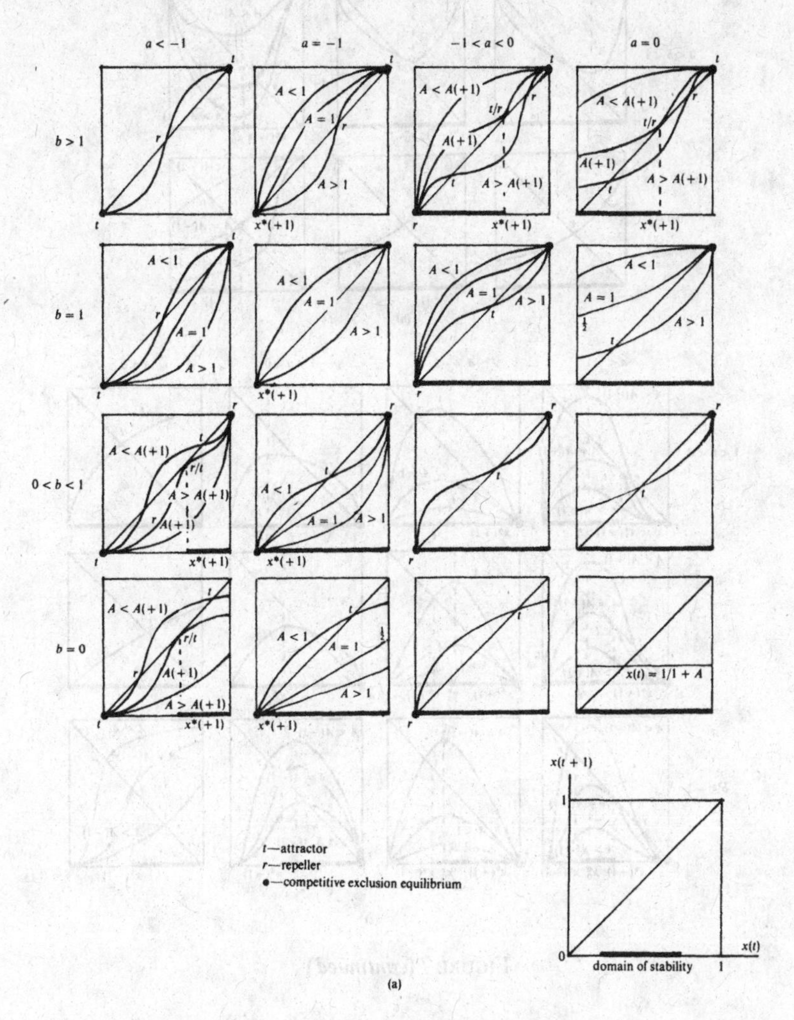

FIGURE 7. (a)–(d) Classification of the 49 cases: different behaviors in the log-linear iterative discrete relative dynamics of the universal map.

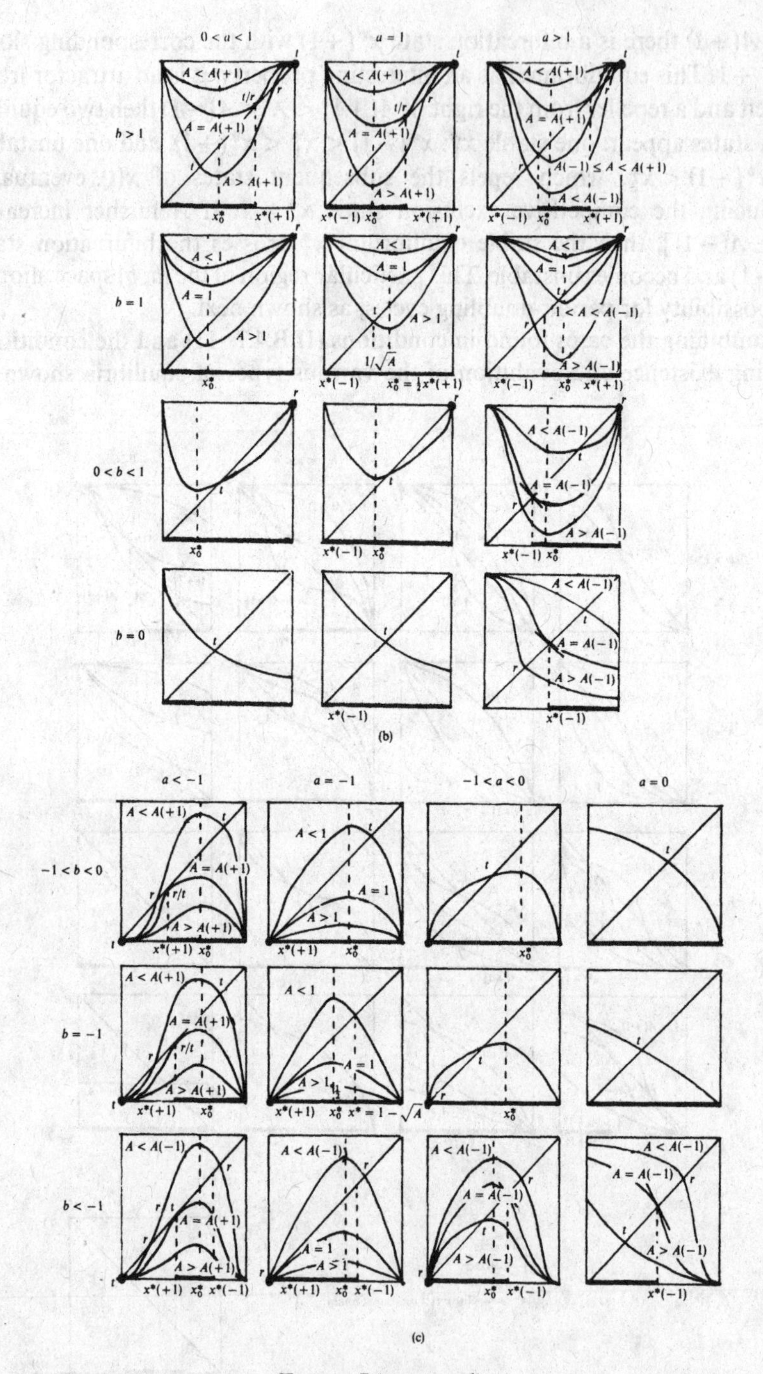

FIGURE 7 (*continued*)

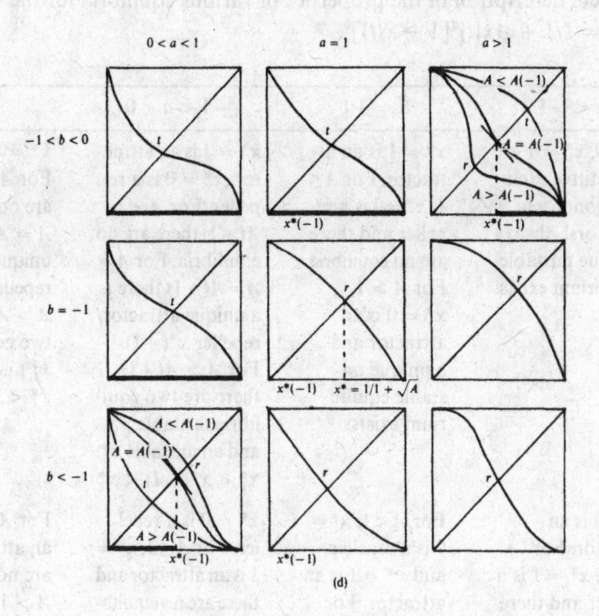

FIGURE 7 (*continued*)

Figure 4, one obtains the comprehensive view of Figure 7. The complete description of the nature and stability of these equilibria are found in Table 1.

Figure 7 demonstrates a symmetry corresponding to a change in the model parameters from (a, b, A) to $(-b, -a, 1/A)$. This symmetry is clearly evident through the substitution $y(t) = 1 - x(t)$

$$y(t+1) = 1 - x(t+1) = 1/1 + \frac{1}{A} x(t)^{-a}[1 - x(t)]^{-b}$$

$$= 1/1 + \frac{1}{A} y(t)^{-b}[1 - y(t)]^{-a}. \qquad \text{(II.B.4.14)}$$

C. Higher Iterates and Fundamental Bifurcations in Discrete Dynamics

1. *The Second Iterate and Two-Period Cycles*

In this section we address the second (and higher) iterate(s) of the log-linear version of the discrete relative one-stock, two-location map (subsection 1), where the three fundamental bifurcations found in these dynamics are presented. We include an analytical description of two-period cycles (subsection 2); the period-doubling process and the Feigenbaum sequence, shown to hold over slopes rather than parameter values, in subsection 3; and the discrete dynamics equivalent of the Hopf bifurcation found in continuous dynamics (subsection 4), the so-called "flip" bifurcation.

TABLE 1. A complete description of the properties of various equilibria for the first iterate of the map: $x(t+1) = 1/1 + Ax(t)^a[1 - x(t)]^b$.

(a)

	$a < -1$	$a = -1$	$-1 < a < 0$	$a = 0$
$b > 1$	$x^* = 0$, $x^* = 1$ are competitive exclusion monotonic attractors. Always a unique unstable equilibrium exists.	$x^* = 1$ is an attractor. For $A \leq 1$, $x^* = 0$ is a repeller and there are no equilibria. For $A > 1$, $x^* = 0$ is an attractor and a unique unstable equilibrium exists.	$x^* = 1$ is an attractor. $x^* = 0$ is a repeller. For $A < A(+1)$ there are no equilibria. For $A = A = A(+1)$ there is a unique attractor/repeller $x^*(+1)$. For $A > A(+1)$ there are two equilibria, a stable x_1^* and an unstable x_2^*: $x_1^* < x^*(+1) < x_2^*$	$x^* = 1$ is an attractor. For $A < A(+1)$ there are no equilibria. For $A = A(+1)$ there is a unique attractor/repeller $x^*(+1)$. For $A > A(+1)$ there are two equilibria, a stable x_1^* and an unstable x_2^*: $x_1^* < x^*(+1) < x_2^*$
$b = 1$	$x^* = 0$ is an attractor. For $A \geq 1$, $x^* = 1$ is a repeller and there are no equilibria. For $A < 1$, $x^* = 1$ is an attractor and a unique unstable equilibria exists.	For $A < 1$, $x^* = 0$ is a repeller and $x^* = 1$ is an attractor. For $A > 1$, $x^* = 0$ is an attractor and $x^* = 1$ is a repeller. In both cases there are no equilibria. For $A = 1$ each state is a fixed point: $x(t+1) = x(t)$.	$x^* = 0$ is a repeller. For $A \leq 1$, $x^* = 1$ is an attractor and there are no equilibria. For $A > 1$, $x^* = 0, 1$ are repellers and a unique stable equilibrium exists.	For $A \leq 1$, $x^* = 1$ is an attractor and there are no equilibria. For $A > 1$, $x^* = 1$ is a repeller and a unique stable equilibrium exists.
$0 < b < 1$	$x^* = 0$ is an attractor and $x^* = 1$ is a repeller. For $A > A(+1)$ there are no equilibria. For $A < A(+1)$ there are two equilibria, a stable x_2^* and an unstable x_1^*: $x_1^* < x^*(+1) < x_2^*$	$x^* = 1$ is repeller. For $A \geq 1$, $x^* = 0$ is an attractor and there are no equilibria. For $A < 1$, $x^* = 0$ is a repeller and a unique stable equilibrium exists.	$x^* = 0, 1$ are repellers and a unique stable equilibrium exists.	$x^* = 1$ is a repeller and a unique stable equilibrium exists.
$b = 0$	$x^* = 0$ is an attractor. For $A > A(+1)$ there are no equilibria. For $A = A(+1)$ a unique repeller/attractor x_1^* exists. For $A < A(+1)$ there are two equilibria, a stable x_2^* and an unstable x_1^*: $x_1^* < x^*(+1) < x_2^*$	For $A \geq 1$, $x^* = 0$ is an attractor and there are no equilibria. For $A < 1$, $x^* = 1$ is a repeller and a unique stable equilibrium exists.	$x^* = 0$ is a repeller and a unique stable equilibrium exists.	$x(t) = 1/1 + A = \text{const.}$

TABLE 1 (*continued*)

(b)

	$0<a<1$	$a=1$	$a>1$
$b>1$	$x^*=0$ is an attractor. For $A<A(+1)$ there are no equilibria. For $A=A(+1)$ a unique attractor/repeller $x^*(+1)$ exists. For $A>A(+1)$ there are two equilibria, a stable x_1^* and an unstable x_2^*: $x_1^*<x^*(+1)<x_2^*$	$x^*=0$ is an attractor. For $A<A(+1)$ there are no equilibria. For $A=A(+1)$ a unique attractor/repeller $x^*(+1)$ exists. For $A>A(+1)$ there are two equilibria, a stable x_1^* and an unstable x_2^*: $0=x^*(-1)<x_1^*<x^*(+1)<x_2^*$	$x^*=0$ is an attractor. For $A<A(+1)$ there are no equilibria. For $A=A(+1)$ a unique attractor/repeller $x^*(+1)$ exists. For $A(+1)<A<A(-1)$ there are two equilibria, a stable x_1^* and an unstable x_2^*: $0<x^*(-1)<x_1^*<x^*(+1)<x_2^*<1$ For $A\geq A(-1)$ both equilibria are unstable: $x_1^*<x^*(-1)<x^*(+1)<x_2^*$
$b=1$	For $A\leq 1$, $x^*=1$ is an attractor and there are no equilibria. For $A>1$ a unique stable equilibrium exists.	For $A\leq 1$, $x^*=1$ is an attractor and there are no equilibria. For $A>1$ a unique stable equilibrium exists.	For $A\leq 1$, $x^*=1$ is an attractor and there are no equilibria. For $1<A<A(-1)$, $x^*=1$ is a repeller and a unique stable equilibrium exists. For $A>A(-1)$ this equilibrium becomes unstable.
$0<b<1$	$x^*=1$ is a repeller and an unique stable equilibrium exists.	$x^*=1$ is a repeller and an unique stable equilibrium exists.	$x^*=1$ is a repeller and a unique equilibrium exists, which is stable for $A<A(-1)$ and becomes unstable for $A\geq A(-1)$.
$b=0$	A unique stable equilibrium exists.	A unique stable equilibrium exists.	A unique equilibrium exists, which is stable for $A<A(-1)$ and becomes unstable for $A\geq A(-1)$.

(c)

	$a<-1$	$a=-1$	$-1<a<0$	$a=0$
$-1<b<0$	$x^*=0$ is an attractor. For $A<A(+1)$ there are no equilibria. For $A=A(+1)$ a unique attractor/repeller $x^*(+1)$ exists. For $A>A(+1)$ there are two equilibria, a stable x_2^* and an unstable x_1^*: $x_1^*<x^*(+1)<x_2^*$	For $A\geq 1$, $x^*(+1)=0$ is an attractor and there are no equilibria. For $A<1$, $x^*(+1)=0$ is a repeller and a unique stable equilibrium exists.	$x^*=0$ is a repeller and a unique stable equilibrium exists.	A unique stable equilibrium exists.
$b=-1$	$x^*=0$ is an attractor. For $A<A(+1)$ there are no equilibria. For $A=A(+1)$ a unique repeller/attractor $x^*(+1)$ exists. For $A>A(+1)$ there are two equilibria, stable x_2^* and unstable x_1^*: $x_1^*<x^*(+1)<x_2^*$	For $A\geq 1$, $x^*(+1)=0$ is an attractor and there are no equilibria. For $A<1$, $x^*(+1)=0$ is a repeller an a unique stable equilibrium exists: $x^*=1-\sqrt{A}$	$x^*=0$ is a repeller and a unique stable equilibrium exists.	A unique stable equilibrium exists: $x^*=\dfrac{A+2-\sqrt{A(A+4)}}{2}$

TABLE 1 (*continued*)

(c)

	$a < -1$	$a = -1$	$-1 < a < 0$	$a = 0$
$b < -1$	$x^* = 0$ is an attractor. For $A > A(+1)$ there are no equilibria. For $A = A(+1)$ a unique repeller/attractor $x^*(+1)$ exists. For $A(-1) < A < A(+1)$ there are two equilibria, stable x_2^* and unstable x_1^*: $x_2^* < x^*(+1) < x_1^* < x^*(-1)$. For $A < A(-1)$ the stable equilibrium x_2^* becomes unstable: $x^*(-1) < x_2^*$.	For $A \geq 1$, $x^*(+1) = 0$ is an attractor and there are no equilibria. For $A(-1) < A < 1$, $x^*(+1)$ is a repeller and a unique stable equilibrium x^* exists: $x^* < x^*(-1)$. For $A < A(-1)$ this equilibrium becomes unstable: $x^*(-1) < x^*$.	$x^* = 0$ is a repeller and a unique equilibrium x^* exists, which is stable for $A > A(-1)$: $x^* < x^*(-1)$, and becomes unstable for $A \leq A(-1)$: $x^* > x^*(-1)$.	A unique equilibrium exists, which is stable for $A > A(-1)$ and becomes unstable for $A \leq A(-1)$

(d)

	$0 < a < 1$	$a = 1$	$a > 1$
$-1 < b < 0$	A unique stable equilibrium exists.	A unique stable equilibrium exists.	A unique equilibrium exists, which is stable for $A < A(-1)$ and becomes unstable for $A(-1) \leq A$.
$b = -1$	A unique stable equilibrium exists.	A unique unstable equilibrium x^* exists: $x^* = 1/1 + \sqrt{A}$	A unique unstable equilibrium exists.
$b < -1$	A unique equilibrium exists, which is stable for $A > A(-1)$ and becomes unstable for $A \leq A(-1)$.	A unique unstable equilibrium x^* exists.	A unique unstable equilibrium exists.

Forming the second iterate $x(t + 2)$ of the log-linear version of our discrete relative map one obtains

$$x(t+2) = 1/1 + Ax(t+1)^a[1 - x(t+1)]^b$$
$$= 1\bigg/\left\{1 + \frac{A^{1+b}x(t)^{ab}[1 - x(t)]^{b^2}}{[1 + Ax(t)^a(1 - x(t))^b]^{a+b}}\right\}. \qquad \text{(II.C.1.1)}$$

The slope $s(t + 2, t)$ of the function $x(t + 2)$ in the $[x(t + 2), x(t)]$ space is, by the chain rule,

$$s(t+2,t) = \frac{\partial x(t+2)}{\partial x(t)} = \frac{\partial x(t+2)}{\partial x(t+1)} \frac{\partial x(t+1)}{\partial x(t)}$$
$$= \frac{1-x(t+2)}{1-x(t)} \frac{x(t+2)}{x(t)} [(a+b)x(t+1)-a][(a+b)x(t)-a]. \tag{II.C.1.2}$$

Fixed points of the second iterate (i.e., $x(t+2) = x(t)$) are the two-period cycles of the first iterate, and they appear in pairs $[x^*(0), x^*(1)]$ so that $x(t) = x^*(0)$, $x(t+1) = x^*(1)$, $x(t+2) = x^*(0)$. A proper two-period cycle appears when $x^*(0) \neq x^*(1)$, a situation which can be captured by the conditions

$$\frac{1-x^*(1)}{x^*(1)} = Ax^*(0)^a[1-x^*(0)]^b,$$
$$\frac{1-x^*(0)}{x(0)} = Ax(1)^a[1-x(1)]^b, \tag{II.C.1.3}$$

or

$$\left(\frac{x^*(0)}{x^*(1)}\right)^{a-1} \left(\frac{1-x^*(0)}{1-x^*(1)}\right)^{b+1} = 1. \tag{II.C.1.4}$$

This equality immediately gives the domain of nonexistence of two-period cycles. By taking the logarithm of the left- and right-hand side of (II.C.1.4) we obtain

$$(a-1)\ln\frac{x^*(0)}{x^*(1)} + (1+b)\ln\frac{1-x^*(0)}{1-x^*(1)} = 0. \tag{II.C.1.5.1}$$

Suppose that $x^*(0) < x^*(1)$, then

$$\frac{x^*(0)}{x^*(1)} < 1 \quad \rightarrow \quad \ln\frac{x^*(0)}{x^*(1)} < 0, \tag{II.C.1.5.2}$$

$$\frac{1-x^*(0)}{1-x^*(1)} > 1 \quad \rightarrow \quad \ln\frac{1-x^*(0)}{1-x^*(1)} > 0, \tag{II.C.1.5.3}$$

and condition (II.C.1.5.1) becomes

$$\frac{1-a}{1+b} = \left[\ln\frac{1-x^*(0)}{1-x^*(1)}\right] \Bigg/ \left[\ln\frac{x^*(0)}{x^*(1)}\right] < 0. \tag{II.C.1.5.4}$$

If $a > 1$ and $b > -1$ a two-period cycle is possible; if $a < 1$ and $b < -1$, a two-period cycle is also feasible. In all other cases we obtain a contradiction given (II.C.1.5.4), Figure 8, with the exception of the point, $a = 1$, $b = -1$, which makes condition (II.C.1.5.1.) hold as an identity. Consequently, the domain of nonexistence of two-period cycles includes the areas given by

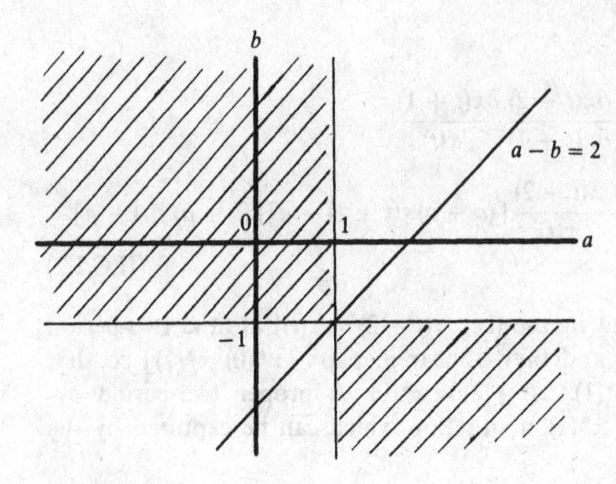

FIGURE 8. A domain (shaded area) of non-existence of two-period cycles for the log-linear model.

conditions

$$(1)\begin{cases}a<1,\\ b>-1,\end{cases}\quad (2)\begin{cases}a>1,\\ b<-1,\end{cases}\quad (3)\begin{cases}a=1,\\ b\neq -1,\end{cases}\quad (4)\begin{cases}a\neq 1,\\ b=-1.\end{cases}\qquad \text{(II.C.1.6)}$$

Thus, the domain of (possible) existence of two-period cycles is

$$(1)\begin{cases}a>1,\\ b>-1,\end{cases}\quad (2)\begin{cases}a<1,\\ b<-1,\end{cases}\quad \text{and (3) the point}\begin{cases}a=1,\\ b=-1.\end{cases}\qquad \text{(II.C.1.7)}$$

The stability properties of two-period cycles are connected with the values of the slope $s(t+2,t)$ at the equilibrium, $x^*(0)$, $x^*(1)$,

$$\begin{aligned}s(t+2,t)|_{x^*(0)} &= s_2^*[x^*(0)] = [(a+b)x^*(0)-a][(a+b)x^*(1)-a]\\ &= s(t+2,t)|_{x^*(1)} = s_2^*[x^*(1)].\end{aligned}\qquad \text{(II.C.1.8)}$$

Analogous to the case of stable (fixed-point) equilibria (one-period cycles), the condition of stability of a two-period cycle, $x^*(0)$, $x^*(1)$, is

$$|s_2^*[x^*(0)]| = |s_2^*[x^*(1)]| < 1.\qquad \text{(II.C.1.9)}$$

The stable two-period cycle case implies that

$$\lim_{t\to\infty} x(t+2) = \lim_{t\to\infty} x(t) = x^*(0) \neq \lim_{t\to\infty} x(t+1) = x^*(1).\qquad \text{(II.C.1.10)}$$

Figure 9 presents the behavior of the log-linear discrete relative dynamics map near the states of stable two-period cycles. The discrete equivalent to a stable limit cycle for continuous dynamics is thus presented.

2. *Analytical Description of Two-Period Cycles*

Analytical descriptions of all two-period cycles are difficult to obtain in general. We try to solve this problem by constructing the analytical algorithm

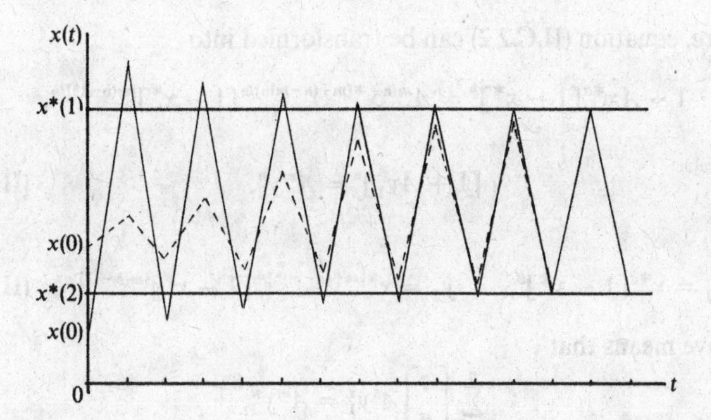

FIGURE 9. A stable two-period cycle: the discrete dynamics equivalent of the stable limit cycle of continuous dynamics.

leading to the complete description of two-period cycles. Expression (II.C.1.1) gives the following equation for any two-period cycle x^*

$$x^* = 1/1 + \frac{A^{b+1}x^{*ab}[1 - x^*]^{b^2}}{[1 + Az^{*a}(1 - x^*]^b]^{a+b}}, \qquad \text{(II.C.2.1)}$$

or

$$1 + Ax^{*a}(1 - x^*)^b = A^{(b+1)/(a+b)}x^{*(ab+1)/(a+b)}(1 - x^*)^{(b^2-1)/(a+b)}. \qquad \text{(II.C.2.2)}$$

The solutions of this nonlinear equation also contain, together with all proper two-period cycles, all equilibrium (nonperiodic) states satisfying the condition (II.B.3.1), i.e., $A = [x^*]^{-(a+1)}[1 - x^*]^{1-b}$.

It is easy to check that conditions (II.C.1.7), which define the domain of possible existence of two-period cycles, are equivalent to

$$0 < \frac{b+1}{a+b} < 1. \qquad \text{(II.C.2.3)}$$

Let us assume that

$$\frac{b+1}{a+b} = \frac{m}{n}. \qquad \text{(II.C.2.4)}$$

where m, n are integers such that $m < n$. This assumption is sufficient to numerically identify the values of two-period cycles. Assumption (II.C.2.4) means that

$$a = \frac{n + (n - m)b}{m}; \qquad \frac{ab + 1}{a + b} = \frac{m + (n - m)b}{n}; \qquad \frac{b^2 - 1}{a + b} = \frac{m(b - 1)}{n}.$$

Therefore, equation (II.C.2.2) can be transformed into

$$1 + Ax^{*a}[1 - x^*]^b = A^{m/n}x^{*[m+(n-m)b]/n}[1 - x^*]^{[m(b-1)]/n},$$

or

$$[1 + Ay_1]^n = A^m y_2^n, \tag{II.C.2.5}$$

where

$$y_1 = x^{*a}[1 - x^*]^b, \qquad y_2 = x^{*[m+(n-m)b]/n}[1 - x^*]^{[m(b-1)]/n}. \tag{II.C.2.6}$$

The above means that

$$\sum_{s=0}^{n} \begin{bmatrix} s \\ n \end{bmatrix} A^s y_1^s = A^m y_2^n,$$

or

$$\rho_n(A) = \sum_{\substack{s=0 \\ s \neq m}}^{n} \begin{bmatrix} s \\ n \end{bmatrix} A^s y_1^s + \left\{ \begin{bmatrix} m \\ n \end{bmatrix} y_1^m - y_2^n \right\} A^m = 0. \tag{II.C.2.7}$$

As mentioned earlier, the solutions of the equation (II.C.2.2), and therefore also the solutions of (II.C.2.7), contain one-period states, satisfying the equation

$$A - y_3 = 0, \tag{II.C.2.8}$$

where

$$y_3 = [x^*]^{-(a+1)}[1 - x^*]^{1-b}. \tag{II.C.2.9}$$

This means that it is possible to state the polynomial $\rho_n(A)$ from (II.C.2.7) in the form

$$\rho_n(A) = (A - y_3)Q_{n-1}(A). \tag{II.C.2.10}$$

Those roots x^* of the polynomial $Q_{n-1}(A)$ which fall between zero and one are the proper two-period cycles of the log-linear discrete relative dynamics.

The structure of the polynomial $Q_{n-1}(A)$ can be found as follows: let

$$Q_{n-1}(A) = \sum_{s=0}^{n-1} A^s z_s, \tag{II.C.2.11}$$

where z_s are unknown coefficients depending on x^*. Then, equality (II.C.2.10) can be written as

$$\sum_{\substack{s=0 \\ s \neq m}}^{n} \begin{bmatrix} s \\ n \end{bmatrix} A^s y_1^s + \left\{ \begin{bmatrix} m \\ n \end{bmatrix} y_1^m - y_2^n \right\} A^m = (A - y_3) \sum_{s=0}^{n-1} A^s z_s \tag{II.C.2.12}$$

$$= \sum_{s=1}^{n-1} [z_{s-1} - y_3 z_s] A^s + z_{n-1} A^n - y_3 z_0.$$

Equating the coefficients of A^s from both sides of equality (II.C.2.12) gives us the coefficients of the polynomial $Q_{n-1}(A)$

$$z_{n-1} = y_1^n,$$

$$z_{n-2} = \begin{bmatrix} 1 \\ n \end{bmatrix} y_1^{n-1} + y_3 y_1^n,$$

$$z_{n-3} = \begin{bmatrix} 2 \\ n \end{bmatrix} y_1^{n-2} + \begin{bmatrix} 1 \\ n \end{bmatrix} y_3 y_1^{n-1} + y_3^2 y_1^n,$$

$$\vdots$$

$$z_{n-(n-m-1)} = z_{m+1} = \begin{bmatrix} m+2 \\ n \end{bmatrix} y_1^{m+2} + \begin{bmatrix} m+1 \\ n \end{bmatrix} y_3 y_1^{m+1} + \cdots + \begin{bmatrix} 1 \\ n \end{bmatrix} y_3^{n-m-3} y_1^{n-1} + y_3^{n-m-2} y_1^n,$$

$$z_m = \frac{1}{y_3^{m+1}} - \begin{bmatrix} 1 \\ n \end{bmatrix} \frac{y_1}{y_3^m} - \cdots - \begin{bmatrix} m \\ n \end{bmatrix} \frac{y_1^m}{y_3},$$

$$\vdots$$

$$z_2 = -\frac{1}{y_3^3} - \begin{bmatrix} 1 \\ n \end{bmatrix} \frac{y_1}{y_3^2} - \begin{bmatrix} 2 \\ n \end{bmatrix} \frac{y_1^2}{y_3},$$

$$z_1 = -\frac{1}{y_3^2} - \begin{bmatrix} 1 \\ n \end{bmatrix} \frac{y_1}{y_3},$$

$$z_0 = -\frac{1}{y_3}. \tag{II.C.2.13}$$

In the above formulas the term y_2 disappears due to condition (II.C.2.9). These formulas identify the coefficients of the equation $Q_{n-1}(A) = 0$, which is the equation containing all proper two-period cycles x^*. As an example we will consider the case (see Figure 8) where

$$\frac{1+b}{a+b} = \frac{1}{2} \qquad \text{or} \qquad a = b+2.$$

In this case, the polynomial $Q_{n-1}(A) = Q_1(A)$ is

$$\begin{aligned} Q_1(A) = z_1 A + z_0 &= y_1^2 A - \frac{1}{y_3} \\ &= \{x^{*b+2}[1-x^*]^b\}^2 A - x^{*b+3}[1-x^*]^{b-1} \\ &= \{x^{*b+2}[1-x^*]^b\}^2 \{A - x^{*-(b+1)}[1-x^*]^{-(b+1)}\}, \end{aligned}$$

and, therefore, the equation for two-period cycles is

$$x^*(1-x^*) = A^{-1/(b+1)},$$

with solutions

$$x^*(0) = \tfrac{1}{2} - \sqrt{\tfrac{1}{4} - A^{-1/(b+1)}}, \qquad x^*(1) = \tfrac{1}{2} + \sqrt{\tfrac{1}{4} - A^{-1/(b+1)}}. \quad \text{(II.C.2.14)}$$

These formulas imply the following (see (II.B.2.5), (II.B.3.8)): for $a = b + 2$ the bifurcation states $x^*(-1)$, corresponding to a slope $s^* = -1$, is

$$x^*(-1) = \frac{a-1}{a+b} = \frac{1}{2},$$

and the corresponding value of the parameter $A = A(-1)$ generating this bifurcation state is

$$A(-1) = [x^*(-1)]^{-(a+1)}[1 - x^*(-1)]^{1-b} = \left[\frac{1}{2}\right]^{-(a+b)} = 4^{1+b}.$$

Existence of proper two-period cycles is guaranteed when the discriminant from (II.C.2.14) is positive

$$\Delta = \tfrac{1}{4} - A^{-1/(b+1)} > 0; \quad \text{(II.C.2.15)}$$

or

$$A > 4^{b+1} = A(-1) \quad \text{if} \quad a > 1, \quad b > -1; \quad \text{(II.C.2.16)}$$

$$A < 4^{b+1} = A(-1) \quad \text{if} \quad a < 1, \quad b < -1. \quad \text{(II.C.2.17)}$$

As A increases from 0 to $+\infty$ in the domain, $a > 1$, $b > -1$, the equilibrium x^* undergoes a successive transition from a fixed-point attractor $[A < A(-1)]$ to a repeller $[A > A(-1)]$, so that this transition is accompanied by the appearance of a two-period cycle $[A > A(-1)]$. The analogical transition also holds in the domain, $a < 1$, $b < -1$, as A decreases from $+\infty$ to zero.

The domain of stability of the two-period cycle, $x^*(0)$, $x^*(1)$, is determined by the inequality $|s_2^*[x^*(0)]| = |s_2^*[x^*(1)]| < 1$. In our case

$$s_2^*[x^*(0)] = s_2^*[x^*(1)] = [(a+b)(\tfrac{1}{2} - \sqrt{\Delta}) - a][(a+b)(\tfrac{1}{2} + \sqrt{\Delta}) - a]$$
$$= 1 - (a+b)^2\Delta = 1 - 4(b+1)^2\Delta. \quad \text{(II.C.2.18)}$$

Therefore, the domain of stability is determined by condition

$$-1 < 1 - 4(b+1)^2\Delta < 1, \quad \text{(II.C.2.19)}$$

or

$$1 - \frac{2}{(b+1)^2} < \frac{4}{A^{1/(b+1)}}. \quad \text{(II.C.2.20)}$$

If $|b + 1| \leq \sqrt{2}$, then this inequality is always true, and conditions (II.C.2.16, 17) give the domain of existence and stability of two-period cycles (II.C.2.14)).

If $|b+1| > \sqrt{2}$, then the domain of stability is given by the inequalities

$$4^{b+1} < A < 4^{b+1}\left[1 - \frac{2}{(b+1)^2}\right]^{-(b+1)} \quad \text{for} \quad b > \sqrt{2} - 1; \qquad \text{(II.C.2.21)}$$

$$4^{b+1}\left[1 - \frac{2}{(b+1)^2}\right]^{-(b+1)} < A < 4^{b+1} \quad \text{for} \quad b < -\sqrt{2} - 1. \qquad \text{(II.C.2.22)}$$

The situation described in this example is typical for two-period cycles: if the parameter A crosses the value $A(-1)$ then stable equilibrium states become unstable and two-period cycles appear. These cycles are at the beginning stable, and after a while they become unstable, giving way to four-, eight-, and in general a succession of 2^n- period cycles, where $n = 1, 2, \ldots$.

3. *Period-Doubling and the Feigenbaum Slope-Sequences*

Consider a quadratic approximation of the Taylor series expansion of the first iterate $x(t+1)$ in a neighborhood of its equilibrium state x^*

$$x(t+1) = x^* + s^*[x(t) - x^*] + \tfrac{1}{2}S^*[x(t) - x^*]^2, \qquad \text{(II.C.3.1)}$$

where

$$s^* = \left.\frac{\partial x(t+1)}{\partial x(t)}\right|_{x^*}, \qquad S^* = \left.\frac{\partial^2 x(t+1)}{\partial x(t)^2}\right|_{x^*}. \qquad \text{(II.C.3.2)}$$

It is possible to find two-period cycles, $x^*(0)$, $x^*(1)$, for this approximative dynamics with the help of the formulas (Sonis and Dendrinos, 1987b)

$$x^*(0) = x^* - \frac{1}{S^*}[1 + s^* - \sqrt{(1+s^*)(s^*-3)}],$$
$$x^*(1) = x^* - \frac{1}{S^*}[1 + s^* + \sqrt{(1+s^*)(s^*-3)}]. \qquad \text{(II.C.3.3)}$$

These solutions are real numbers if and only if $(1+s^*)(s^*-3) > 0$; therefore, the conditions for existence of two-period cycles are

$$s^* < -1 \quad \text{or} \quad s^* > 3. \qquad \text{(II.C.3.4)}$$

The domains $s^* < 1$ and $s^* > 3$ are symmetrical with respect to the point $s^* = +1$, and each of them can be transformed into the other by the substitution

$$\tilde{s}^* = 2 - s^*. \qquad \text{(II.C.3.5)}$$

Therefore each solution (II.C.3.3) at any value of the slope s^* generates a "shifted" solution at another slope value $\tilde{s}^* = 2 - s^*$ which corresponds to suitable $\tilde{x}^*$ and $\tilde{S}^*$; thus solutions (II.C.3.3) appear in pairs in the domains, $s^* < -1, s^* > 3$. As a result, one can consider only the domain, $s^* < -1$, and obtain corresponding statements for $s^* > 3$ automatically.

The qualitative picture which ensues from the existence of approximative two-period cycles is as follows: if the parameter A crosses the value $A(-1)$ corresponding to the slope $s^*(-1)$ where $s^* = -1$, i.e., if the equilibrium state x^* undergoes a transition from a stable fixed point (attractor) to an unstable fixed point (repeller), then a two-period cycle emerges. The same situation holds for $s^* > 3$. At the beginning, this two-period cycle is stable and beyond some critical value of the slope s^* it becomes unstable. To find the domain of stability of this cycle the presentation of the slope s_2^* as a function of the slope s^* will be derived approximately but analytically (see Helleman, 1981). This presentation can be extracted from the approximation (II.C.3.1). It is possible to check using (II.C.3.1) that under the substitutions $x(t) = x^*(0)$, $x(t+1) = x^*(1)$ and $x(t) = x^*(1)$, $x(t+1) = x^*(0)$

$$x(t+1) - x^*(1) = \{s^* + S^*[x^*(0) - x^*]\}[x(t) - x^*(0)] + \tfrac{1}{2}S^*[x(t) - x^*(0)]^2, \tag{II.C.3.6.a}$$

$$x(t+1) - x^*(0) = \{s^* + S^*[x^*(1) - x^*]\}[x(t) - x^*(1)] + \tfrac{1}{2}S^*[x(t) - x^*(1)]^2. \tag{II.C.3.6.b}$$

Using (II.C.3.6.a, b) one obtains

$$\begin{aligned} x(t+2) - x^*(1) &= \{s^* + S^*[x^*(0) - x^*]\}[x(t+1) - x^*(0)] + \tfrac{1}{2}S^*[x(t+1) - x^*(0)]^2 \\ &= \{s^* + S^*[x^*(0) - x^*]\}\{s^* + S^*[x^*(1) - x^*]\}[x(t) - x^*(1)] \\ &\quad + \tfrac{1}{2}S^*\{[s^* + S^*(x^*(0) - x^*)] \\ &\quad + [s^* + S^*(x^*(1) - x^*)]^2\}[x(t) - x^*(1)]^2 \\ &\quad + \cdots \text{ higher-order terms.} \end{aligned}$$

Analogously

$$\begin{aligned} x(t+2) - x^*(0) &= \{s^* + S^*[x^*(0) - x^*]\}\{s^* + S^*[x^*(1) - x^*]\}[x(t) - x^*(0)] \\ &\quad + \tfrac{1}{2}S^*\{[s^* + S^*(x^*(1) - x^*)] \\ &\quad + [s^* + S^*(x^*(0) - x^*)]^2\}[x(t) - x^*(0)]^2 \\ &\quad + \cdots \text{ higher-order terms.} \end{aligned}$$

Therefore, the expression $\{s^* + S^*[x^*(0) - x^*]\}\{s^* + S^*[x^*(1) - x^*]\}$ is an approximation of the slope s_2^*, i.e., s_2^* can be presented as a function of s^*

$$\begin{aligned} s_2^* &= \{s^* + S^*[x^*(0) - x^*]\}\{s^* + S^*[x^*(1) - x^*]\} \\ &= \{s^* - [1 + s^* - \sqrt{(1+s^*)(s^*-3)}]\}\{s^* - [1 + s^* + \sqrt{(1+s^*)(s^*-3)}]\} \\ &= 1 - (1+s^*)(s^*-3) = 4 + 2s^* - s^{*2}, \end{aligned}$$

and the domain of stability of two-period cycles, $x^*(0), x^*(1)$, is defined by the inequality

$$-1 < 4 + 2s^* - s^{*2} < 1. \qquad \text{(II.C.3.7)}$$

Geometrically, this domain presents, in the case that $s^* < -1$, an interval $[c_-(2), c_-(1)]$ with the end points, $c_-(2), c_-(1)$, corresponding to the negative solutions of two quadratic equations

$$4 + 2s^* - s^{*2} = \pm 1. \qquad \text{(II.C.3.8)}$$

Therefore, for $s^* < -1$ this interval $[c_-(2), c_-(1)]$ is

$$c_-(2) = -1.4495\ldots \simeq 1 - \sqrt{6} < s^* < -1 = c_-(1). \qquad \text{(II.C.3.9)}$$

Obviously, $c_-(2)$ is a negative solution of the equation

$$4 + 2c_-(2) - c_-(2)^2 = c_-(1). \qquad \text{(II.C.3.10)}$$

In the case $s^* > 3$ the interval $[c_+(1), c_+(2)]$ of stability of the two-period cycle is defined by substitution (II.C.3.5)

$$c_+(1) = 3 < s^* < 1 + \sqrt{6} \simeq 3.4495\ldots = c_+(2). \qquad \text{(II.C.3.11)}$$

These results give the basis for the approximate *renormalization* procedure, which leads to the approximative description of the phenomenon of period doubling (Helleman, 1981; Sonis and Dendrinos, 1987b). The meaning of the renormalization procedure lies in the fact that the transfer from the n-iterate to the $(n+1)$-iterate can be presented in (analogical) analytical form with "renormalized" coefficients depending only on n.

If $x^*_{2^n}$ is a fixed point for the nth iterate $x(t+n)$, i.e., $x^*_{2^n}$ is a 2^n-period cycle of $x(t)$, then the approximate description of the dynamics of $x(t+n+1)$ in the vicinity of the 2^n-period cycle $x^*_{2^n}$ is

$$x(t+n+1) = x^*_{2^n} + s^*_{2^n}[x(t+n) - x^*_{2^n}] + \tfrac{1}{2}S^*_{2^n}[x(t+n) - x^*_{2^n}]^2, \qquad \text{(II.C.3.12)}$$

where

$$s^*_{2^n} = \left.\frac{\partial x(t+n)}{\partial x(t)}\right|_{x^*_{2^n}}, \qquad S^*_{2^n} = \left.\frac{\partial^2 x(t+k)}{\partial x(t)^2}\right|_{x^*_{2^n}}, \qquad \text{(II.C.3.13)}$$

and the slope $s^*_{2^n}$ depends on n.

Since the transfer from $x(t+n+1)$ to $x(t+n)$ is analogical to an analytical transfer from $x(t+1)$ to $x(t)$ one can obtain the formula for the 2^{n+1}-period cycle (analogical to (II.C.3.3))

$$x^*_{2^{n-1}} = x^*_{2^n} - \frac{1 + s^*_{2^n} \pm \sqrt{[(1 + s^*_{2^n})(s^*_{2^n} - 3)}}{S^*_{2^n}}, \qquad \text{(II.C.3.14)}$$

provided that $s^*_{2^n} < -1$ or $s^*_{2^n} > 3$. In the case where $s^*_{2^n} < -1$, the negative solutions of the recurrent equation (see (II.C.3.10))

$$4 + 2c_-(2^{n+1}) - c_-(2^{n+1})^2 = c_-(2^n), \qquad c_-(1) = -1, \qquad \text{(II.C.3.15)}$$

give the points of the interval $[c_-(2^{n+1}), c_-(2^n)]$ of stability of the 2^n-period cycle

$$c_-(2^{n+1}) < s^* < c_-(2^n), \tag{II.C.3.16}$$

where s^* is *the slope* of the discrete map for $x(t+1)$. The sequence $c_-(2^n)$ will be called the *approximative negative Feigenbaum slope-sequence.* Its entries, i.e., the bifurcation points $c_-(2^n)$, can be calculated from (II.C.3.15)

$$c_-(2^{n+1}) = 1 - \sqrt{5 - c_-(2^n)}, \qquad c_-(1) = -1, \tag{II.C.3.17}$$

so that the negative slope-sequence is

$$-1, -1.4495\ldots, -1.5396\ldots, -1.5573\ldots, -1.5607\ldots, -1.5614\ldots. \tag{II.C.3.18}$$

For $s^*_{2^n} > 3$ one can obtain, by a substitution from (II.C.3.5), $c_+(2^n) = 2 - c_-(2^n)$, the *approximative positive Feigenbaum slope-sequence*

$$3, 3.4495\ldots, 3.5396\ldots, 3.5573\ldots, 3.5607\ldots, 3.5614\ldots. \tag{II.C.3.19}$$

This sequence is generated by the recurrent formula

$$c_+(2^{n+1}) = 1 + \sqrt{3 + c_+(2^n)}, \qquad c_+(1) = 3. \tag{II.C.3.20}$$

Initially, Feigenbaum (1978) constructed the sequence (II.C.3.19) for the values of the structural *parameter*, a, in the logistic difference equation: $x(t+1) = ax(t)[1 - x(t)]$, $a \leq 4$. For this equation it so happens that there is a linear dependence between the parameter, a, and the slope, s^*, of the discrete map, $s^* = 2 - a$. For difference equations more complicated than the logistic equation, the connection between parameters and slopes is arbitrarily complex, and thus it is impossible to find, in general, universal sequences for parameter values. In a capsule, our aforementioned arguments prove that the *universality of Feigenbaum sequences is related not to the parameters but to the slopes of the discrete map* (!). This issue is more fully discussed in the paper by Sonis and Dendrinos (1987b).

Further, the sequence of bifurcation points $c_-(2^n)$ tends to the limit

$$\lim_{n\to\infty} = c_-(2^n) = c_-,$$

which is the fixed point of the transformation (II.C.3.17); therefore

$$c_- = 1 - \sqrt{5 - c_-} \qquad \text{or} \qquad c_- = \frac{1 - \sqrt{17}}{2} \approx -1.5616\ldots. \tag{II.C.3.21}$$

The asymptotic behavior of the negative Feigenbaum slope sequence is characterized by the limit

$$\lim_{n\to\infty} \frac{c_-(2^{n+1}) - c_-}{c_-(2^n) - c_-} = \lim_{n\to\infty} \frac{c_-(2^{n+1}) - c_-(2^n)}{c_-(2^n) - c_-(2^{n-1})} = \delta_-. \tag{II.C.3.22}$$

The other form of (II.C.3.22) is

$$c_-(2^n) \underset{n\to\infty}{\sim} c_- + [c_-(1) - c_-]\delta_-^n.$$

Substituting this into (II.C.3.15) one obtains

$$\begin{aligned} &c_- + [c_-(1) - c_-]\delta_-^n \\ &\quad = 4 + 2\{c_- + [c_-(1) - c_-]\delta^{n+1}\} - \{c_- + [c_-(1) - c_-]\delta^{n+1}\}^2 \\ &\quad \approx c_- + 2[c_-(1) - c_-](1 - c_-)\delta_-^{n+1} + \text{higher-order terms.} \end{aligned}$$

This implies that

$$\delta_- \approx 1/2(1 - c_-) = 1/1 + \sqrt{17} \approx 0.1952\ldots, \qquad \text{(II.C.3.23)}$$

and the asymptotical behavior of the negative Feigenbaum slope-sequence is now

$$c_-(2^n) \underset{n\to\infty}{\sim} -1.5616\ldots + 2.5616\ldots(0.1952\ldots)^n. \qquad \text{(II.C.3.24)}$$

Thus, the negative slope-sequence converges geometrically to the final bifurcation point c_- with a constant rate of convergence δ_-. After this value of slope $s^* = c_-$, all 2^n-period cycles created have turned unstable, and ultimately chaotic behavior results. The asymptotic behavior of the positive Feigenbaum slope-sequence is easily obtained from the asymptotic behavior of negative slope-sequence by the substitution (II.C.3.5). Here

$$\lim_{n\to\infty} c_+(2^n) = c_+ = 2 - c_- = \frac{3 + \sqrt{17}}{2} \approx 3.56162\ldots, \qquad \text{(II.C.3.25)}$$

$$\begin{aligned} \delta_+ &= \lim_{n\to\infty} \frac{c_+(2^{n+1}) - c_+(2^n)}{c_+(2^n) - c_+(2^{n-1})} \approx 1/2[1 - (2 - c_+)] \\ &= 1/1 + \sqrt{17} = \delta_- = 0.1952\ldots, \end{aligned} \qquad \text{(II.C.3.26)}$$

which gives the following asymptotic behavior

$$\begin{aligned} c_+(2^n) &\underset{n\to\infty}{\sim} c_+ + [c_+(1) - c_+]\delta_+^n \qquad \text{(II.C.3.27)} \\ &= 3.5616\ldots - 0.5616\ldots(0.1952\ldots)^n. \end{aligned}$$

4. *Domains of Nonexistence of k-Period Cycles* ($k \geq 3$) *and the Hopf Equivalent Bifurcation*

The specific form of the discrete map elaborated on so far, puts strong restrictions on the geometrical form of k-period cycles in our log-linear dynamics, since, as was shown in (II.B.4), the discrete map attains only one singular point (see Figure 7). Each k-period cycle

$$x^*(0), x^*(1), \ldots, x^*(k-1), x^*(k) = x^*(0) \qquad \text{(II.C.4.1)}$$

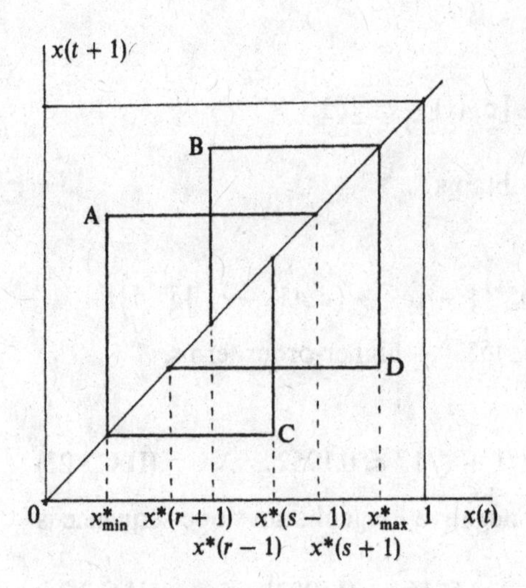

FIGURE 10. The impossible form of k-period cycles.

can be represented geometrically on a discrete map as a closed loop with $2k$ vertices: k of them are on the diagonal of the discrete map. Consider the particular vertices where

$$x^*(s) = x^*_{min} = \min\{x^*(0), x^*(1), \ldots, x^*(k-1)\},$$
$$x^*(r) = x^*_{max} = \max\{x^*(0), x^*(1), \ldots, x^*(k-1)\}. \tag{II.C.4.2}$$

One can prove now that either

$$x^*(s) = x^*_{min}, \qquad x^*(s+1) = x^*_{max}; \tag{II.C.4.3}$$

or

$$x^*(r) = x^*_{max}, \qquad x^*(r+1) = x^*_{min}. \tag{II.C.4.4}$$

If both conditions (II.C.4.3, 4) do not hold then the closed loop of a k-period cycle cannot include vertices (x^*_{min}, x^*_{max}) or (x^*_{max}, x^*_{min}) (see Figure 10), and instead it must include four points

$$A[x^*_{min}, x^*(s+1)], \qquad B[x^*(r-1), x^*_{max}],$$
$$C[x^*(s-1), x^*_{min}], \qquad D[x^*_{max}, x^*(r+1)].$$

This implies that the discrete map includes more than one point of maxima or minima, which is impossible.

Conditions (II.C.4.3, 4) also can be employed to construct the *domain of nonexistence of k-period cycles* ($k \geq 3$). This domain is the intersection of two domains

$$a \leq 1, \qquad b \geq -1, \tag{II.C.4.5}$$

and

$$a \geq 0, \qquad b \leq 0. \tag{II.C.4.6}$$

Let us demonstrate this for the case (II.C.4.3) where $x^*(s) = x^*_{\min}$, $x^*(s+1) = x^*_{\max}$. Condition (II.A.2.15) implies that

$$\frac{1 - x^*_{\min}}{x^*_{\min}} = A[x^*(s-1)]^a[1 - x^*(s-1)]^b, \tag{II.C.4.7}$$

$$\frac{1 - x^*_{\max}}{x^*_{\max}} = A[x^*_{\min}]^a[1 - x^*_{\min}]^b, \tag{II.C.4.8}$$

$$\frac{1 - x^*(s+2)}{x^*(s+2)} = A[x^*_{\max}]^a[1 - x^*_{\max}]^b. \tag{II.C.4.9}$$

Conditions (II.C.4.7, 8) give (by division)

$$\left(\frac{1 - x^*_{\min}}{1 - x^*_{\max}}\right)\frac{x^*_{\max}}{x^*_{\min}} = \left[\frac{x^*(s-1)}{x^*_{\min}}\right]^a\left[\frac{1 - x^*(s-1)}{1 - x^*_{\min}}\right]^b,$$

or

$$\left[\frac{1 - x^*(s-1)}{1 - x^*_{\max}}\right]\frac{x^*_{\max}}{x^*(s-1)} = \left[\frac{x^*(s-1)}{x^*_{\min}}\right]^{a-1}\left[\frac{1 - x^*(s-1)}{1 - x^*_{\min}}\right]^{b+1},$$

or

$$\log\left[\frac{1 - x^*(s-1)}{1 - x^*_{\max}}\right] + \log\left[\frac{x^*_{\max}}{x^*(s-1)}\right] = (a-1)\log\left[\frac{x^*(s-1)}{x^*_{\min}}\right] + (b+1)\log\left[\frac{1 - x^*(s-1)}{1 - x^*_{\min}}\right]. \tag{II.C.4.10}$$

Since $x^*_{\min} < x^*(s-1), x^*(s+2) < x^*_{\max}$ then

$$\frac{1 - x^*(s-1)}{1 - x^*_{\max}} > 1 \quad \rightarrow \quad \log\left[\frac{1 - x^*(s-1)}{1 - x^*_{\max}}\right] > 0,$$

$$\frac{x^*_{\max}}{x^*(s-1)} > 1 \quad \rightarrow \quad \log\left[\frac{x^*_{\max}}{x^*(s-1)}\right] > 0,$$

$$\frac{x^*(s-1)}{x^*_{\min}} < 1 \quad \rightarrow \quad \log\left[\frac{x^*(s-1)}{x^*_{\min}}\right] < 0,$$

$$\frac{1 - x^*(s-1)}{1 - x^*_{\min}} > 1 \quad \rightarrow \quad \log\left[\frac{1 - x^*(s-1)}{1 - x^*_{\min}}\right] > 0.$$

Therefore, if $a \leq 1$, $b \geq -1$, then the left-hand side of the equality (II.C.4.10) will be strictly positive, just as the right-hand side will be negative or zero, and we have a contradiction. Further, conditions (II.C.4.8, 9) imply

$$\frac{1 - x^*_{\max}}{1 - x^*(s+2)}\,\frac{x^*(s+2)}{x^*_{\max}} = \left[\frac{x^*_{\min}}{x^*_{\max}}\right]^2\left[\frac{1 - x^*_{\min}}{1 - x^*_{\max}}\right]^b;$$

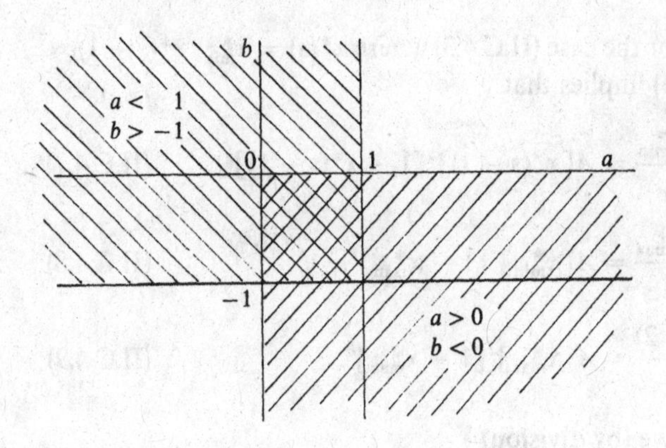

FIGURE 11. Domain of nonexistence of k-period cycles ($k \geq 3$) for the log-linear model.

or

$$\log\left[\frac{1 - x^*_{\max}}{1 - x^*(s+2)}\right] + \log\left[\frac{x^*(s+2)}{x^*_{\max}}\right]$$
$$= a \log\left[\frac{x^*_{\min}}{x^*_{\max}}\right] + b \log\left[\frac{1 - x^*_{\min}}{1 - x^*_{\max}}\right],$$

which in combination with conditions, $a \geq 0$, $b \leq 0$, lead to a contradiction.

Analogously, condition (II.C.4.4) gives the same domain of nonexistence of k-period cycles ($k \geq 3$). Figure 11 shows the domain of nonexistence of k-period cycles (shaded areas); the domain of possible existence of k-period cycles ($k \geq 3$) includes two areas

$$a > 1, \quad b > 0 \qquad \text{and} \qquad a < 0, \quad b < -1. \tag{II.C.4.11}$$

Comparing Figure 11 with Figure 8, the latter presenting the domain of nonexistence of two-period cycles, gives the two areas which can include only two-period cycles (see Figure 12)

$$a > 1, \quad -1 < b < 0 \qquad \text{and} \qquad 0 < a < 1, \quad b < -1. \tag{II.C.4.12}$$

This event strongly resembles the phenomenon of the two-dimensional Hopf

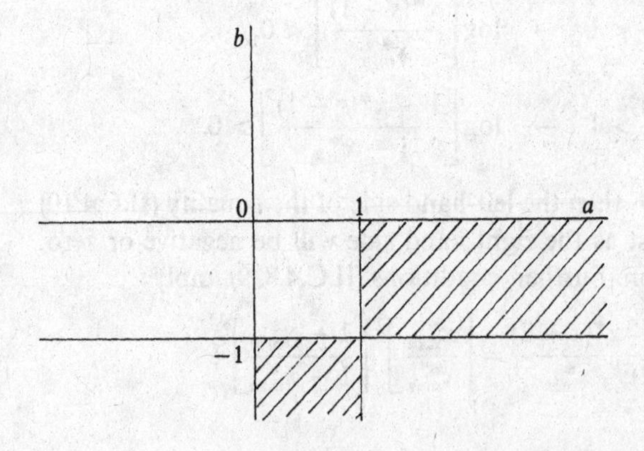

FIGURE 12. Domain of existence of the discrete Hopf bifurcation.

bifurcation for continuous dynamics which switches stable foci, through centers, to stable limit cycles (Marsden and McCraken, 1976). How the Hopf bifurcation operates in discrete dynamics when higher dimensions are involved is demonstrated by Guckenheimer and Holmes (1983) and Iooss *et al.* (1983), among others. However, this universal map sets the stage not only for obtaining a detailed look at how the discrete dynamics Hopf equivalent bifurcation works at the two-dimensional, one-time delay iterative map, but further on how it evolves from two to higher-order dimensions.

In summary, there are three fundamental bifurcations in discrete relative dynamics occurring as the bifurcation parameter A moves smoothly from zero to $+\infty$, or from $+\infty$ to zero:

bifurcation type one: this phenomenon identifies a phase transition where a fixed point, through a center, is switched into a stable two-period cycle: this is the discrete equivalent of the Hopf bifurcation found in continuous dynamics (Figure 13), or the "flip" bifurcation;

bifurcation type two: an unstable equilibrium (competitive exclusion) switches (through a one-sided stable fixed point) to a stable equilibrium;

bifurcation type three: an unstable equilibrium switches to a stable point, and then to a series of period-doubling cycles leading to deterministic chaos.

The map's bifurcating behavior leads us to conclude that the original May turbulence, found in the logistic model $x(t+1) = ax(t)[1 - x(t)]$ when $0 \leq a \leq 4$ and where $0 \leq x(t) \leq 1$, occurs as a result of the conditions responsible

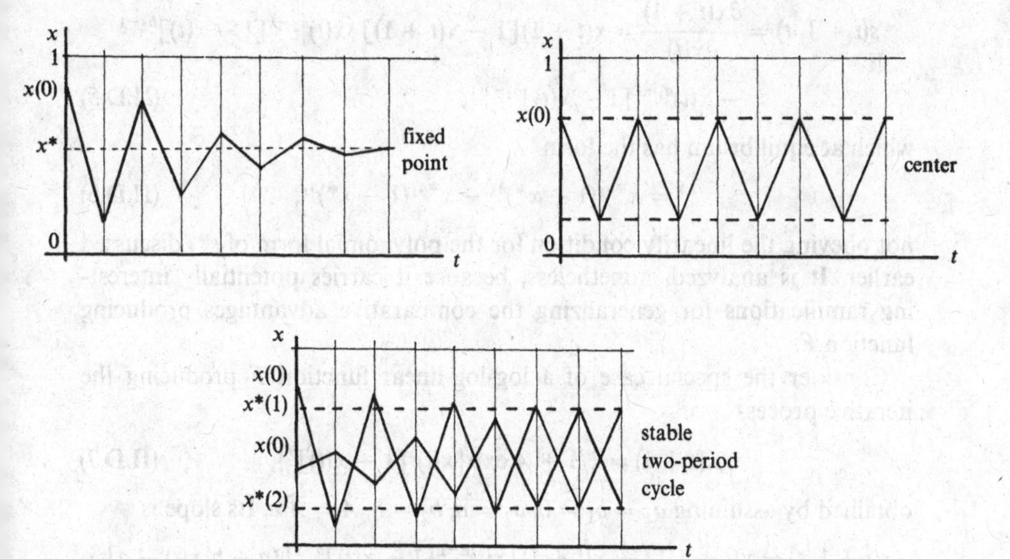

FIGURE 13. The discrete dynamics equivalent of the Hopf bifurcation found in continuous dynamics.

for bifurcation type one and bifurcation type two being met simultaneously at particular domains in the parameter space; specifically when $a, b \geq 1$ and/or $a, b \leq -1$. We have demonstrated this in the paper by Dendrinos and Sonis (1987a).

D. The Exponential Locational Advantages Producing Function

We now consider different F functions, namely the exponential locational advantages production functions

$$F_1[x_1(t), x_2(t)] = \exp[x_1(t)^{a_1} x_2(t)^{b_1}], \tag{II.D.1}$$

$$F_2[x_1(t), x_2(t)] = \exp[x_1(t)^{a_2} x_2(t)^{b_2}], \tag{II.D.2}$$

which generate the iterative process

$$x_1(t+1) = \frac{A_1 \exp[x_1(t)^{a_1} x_2(t)^{b_1}]}{A_1 \exp[x_1(t)^{a_1} x_2(t)^{b_1}] + A_2 \exp[x_1(t)^{a_2} x_2(t)^{b_2}]}, \tag{II.D.3}$$

$$x_2(t+1) = 1 - x_1(t+1), \qquad A_1, A_2 > 0.$$

Denoting $x_1(t) = x(t)$, $A_2/A_1 = \tilde{A}$, condition (II.D.3) obtains the form

$$x(t+1) = 1/1 + \tilde{A} \exp\{x(t)^{a_2}[1 - x(t)]^{b_2} - x(t)^{a_1}[1 - x(t)]^{b_1}\}. \tag{II.D.4}$$

The slope of this iterative process is now given by

$$s(t+1, t) = \frac{\partial x(t+1)}{\partial x(t)} = x(t+1)[1 - x(t+1)]\{x(t)^{a_2 - 1}[1 - x(t)]^{b_2 - 1} - x(t)^{a_1 - 1}[1 - x(t)]^{b_1 - 1}\}, \tag{II.D.5}$$

which at equilibrium has the form

$$s^* = x^{*a_2}(1 - x^*)^{b_2} - x^{*a_1}(1 - x^*)^{b_1}, \tag{II.D.6}$$

not obeying the linearity condition (or the polynomial form of s^*) discussed earlier. It is analyzed, nonetheless, because it carries potentially interesting ramifications for generalizing the comparative advantages producing function F.

Consider the special case of a log-log-linear function F producing the iterative process

$$x(t+1) = 1/1 + A \exp\{x(t)^a[1 - x(t)]^b\}, \tag{II.D.7}$$

obtained by assuming $a_1 = b_1 = 0$, $a_2 = a$, $b_2 = b$, $A = \tilde{A}/e$. Its slope is

$$s(t+1, t) = x(t+1)[1 - x(t+1)]x(t)^{a-1}[1 - x(t)]^{b-1}[(a+b)x(t) - a], \tag{II.D.8}$$

so that at equilibrium

$$s^* = x^{*a}(1 - x^*)^b[(a + b)x^* - a], \tag{II.D.9}$$

$$\frac{1 - x^*}{x^*} = A \exp[x^{*a}(1 - x^*)^b], \tag{II.D.10}$$

$$A = \Phi(x^*) = \frac{1 - x^*}{x^* \exp[x^{*a}(1 - x^*)^b]}. \tag{II.D.11}$$

Expression (II.D.9) implies that

$$\frac{\partial s^*}{\partial x^*} = -x^{*a-1}(1 - x^*)^{b-1}[x^{*2}(a + b)(a + b + 1) - x^*(a + b)(2a + 1) + a^2],$$

meaning that the slope-function $s^*(x^*)$ has the singular points

$$x^*_{1,2} = \frac{2a + 1 \pm \sqrt{1 + 4ab/(a + b)}}{2(a + b + 1)}, \tag{II.D.12}$$

where a maximum/minimum/inflection point in $s^*(x^*)$ is attained. Of particular interest are the points in the parameter space (a, b) where bifurcations occur, where $|s^*| = 1$, whereas the domain of stability is $|s^*| < 1$. From (II.D.9)

$$\lim_{x^* \to 0+} s^* = \begin{cases} 0, & a \geq 0, \\ +\infty, & a < 0, \end{cases} \qquad \lim_{x^* \to 1-} s^* = \begin{cases} 0, & b \geq 0, \\ -\infty, & b < 0, \end{cases} \tag{II.D.13}$$

identifying the area $a, b \geq 0$ as an area of stability, where fixed-point behavior occurs in x^*.

Two-period cycles take place in the $b < 0$ space. This is demonstrated as follows: conditions where two-period cycles occur, $x(t) = x(t + 2) = x^*(0)$, $x(t + 1) = x^*(1)$, $x^*(0) \neq x^*(1)$, imply

$$\begin{aligned} \frac{1 - x^*(1)}{x^*(1)} &= A \exp\{x^*(0)^a[1 - x^*(0)]^b\}, \\ \frac{1 - x^*(0)}{x^*(0)} &= A \exp\{x^*(1)^a[1 - x^*(1)]^b\}. \end{aligned} \tag{II.D.14}$$

Suppose that $x^*(0) < x^*(1)$, then

$$\frac{x^*(0)}{x^*(1)} < 1 \quad \text{and} \quad \frac{1 - x^*(1)}{1 - x^*(0)} < 1$$

and therefore, from (II.D.14)

$$\left[\frac{x^*(0)}{x^*(1)}\right]^a \left[\frac{1 - x^*(0)}{1 - x^*(1)}\right]^b < 1, \tag{II.D.15}$$

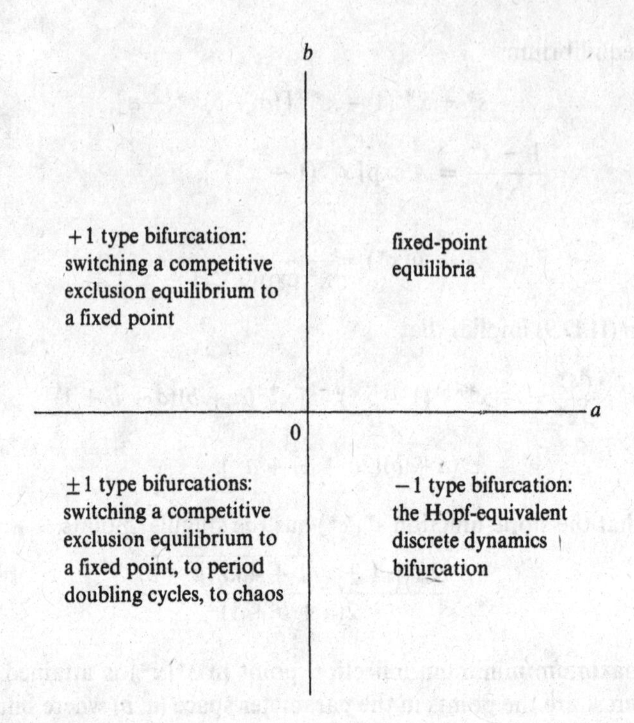

FIGURE 14. Bifurcation types and their domains found in the exponential specifications of the universal map.

or

$$a \ln \frac{x^*(0)}{x^*(1)} + b \ln \frac{1 - x^*(0)}{1 - x^*(1)} < 0. \tag{II.D.16}$$

Since

$$\frac{x^*(0)}{x^*(1)} < 0 \quad \text{and} \quad \ln \frac{1 - x^*(0)}{1 - x^*(1)} > 0,$$

then if $a < 0$ and $b > 0$ we obtain a contradiction. Thus, in the domain $b > 0$, there is no two-period cycle present (Figure 14).

The domain where k (≥ 3)-period cycles occur is that of $a, b < 0$, whereas the Hopf equivalent (flip) discrete dynamics bifurcation ($s_2^* = 1$) occurs in the domain ($a > 0$, $b < 0$). Finally, the $s_1^* = -1$ bifurcation, switching a competitive exclusion equilibrium to a fixed point as the environmental parameter, A, changes in the $(0, +\infty)$ space occurs in ($a < 0$, $b > 0$).

Conclusions

In this part we presented the simplest version of our universal discrete-time relative dynamics model, namely the one-stock, two-region dynamics. A log-linear and an exponential specification of our F functions were analyzed in more detail, analytically and through numerical simulations.

We demonstrated that a number of innovative dynamic phenomena are embedded in our model, not previously discussed in the literature. These results were obtained under the log-linear and exponential specifications of our F functions, drawing their interpretation through references to temporal and comparative advantages in location theory. Among the innovative dynamic phenomena emerging from the discrete map one can include the identification of three fundamental discrete relative dynamics bifurcations. One of these phase transitions is the two-dimensional discrete dynamics equivalent of the Hopf bifurcation found is continuous dynamics. Also, the conditions for the presence of the May type turbulence (one of the three bifurcations) were identified as the simultaneous presence of the other two in the particular region of the parameter space where it appears.

Another major funding of this part was that the Feigenbaum sequences, previously identified over ranges of parameter values, are indeed slope-sequences of the universal one-stock, two-region model. They were found to hold in the case of the log-linear specification of our universal map, too.

Given the variety of dynamic qualitative events built into the specifications of the universal map, the possibility of addressing a host of socio-spatial developmental and evolutionary events was presented. Further, amplified by the fundings of the next two parts, one is assisted in laying the ground work for a structural approach to socio-spatial dynamics.

Our model's state variables, structural and environmental parameters, and the algorithm's form depict the end bundle (composite) effects and results of a complex interplay of social forces extended in time and space. Interpretation of the structural parameters is related to composite temporal, locational, and comparative advantages and associated locationally specific elasticities of growth/decline of stocks.

It was demonstrated that, in this discrete relative dynamics map, a far more complicated catalogue of locational associations are embedded than the simple six ecological associations found in the Volterra–Lotka formalism of continuous dynamics. In all, 49 different associations exist in this one-stock, two-location model.

Part III
One Stock, Multiple Locations

Summary

We now present a generalization of the discrete-time, universal, one-species, two-location, relative dynamics model to I locations. First, the general formulation will be analyzed in Section A; then a log-linear specification of the comparative advantages F function will be studied in Section B, followed by numerical simulations and some general results. One of these results is the novel funding that the onset of local and partial period-doubling cycles may occur in the dynamic system.

However, when turbulence occurs it is much more likely that it is of a global nature, meaning that turbulence in the dynamics of stock(s) occurs most likely at all locations and is not confined to a subset of them. When local turbulence take place, it is of a special type. Some symmetry is detected in the unfolding over space of fixed points, and in stable even-period cycles, when they coexist. Existence of "strange containers and attractors," "state-switching," and associated events are also demonstrated to occur in the catalogue of dynamic events embedded in the universal map. In this part, the empirical evidence available to date is included in Section C. It addresses the relative population dynamics of the U.S. regions in the period 1850–1980 with projections to the year 2050. The empirical work affords the opportunity to link the qualitative dynamic properties of the model to iterative step size, to least squares regression tests, and thus to explore deeper issues of verification, model validation, and testing of socio-spatial hypotheses.

Attempts toward identifying the population size limits of the strange contaniners and attractors are made in Section D; we demonstrate that our universal map can be modified so that it can produce a discrete-time equivalent of the continuous-time logistic growth model, including the standard multinomial logistic growth case (Section E). This extension's stability properties of the dynamic (globally or locally competitive exclusion type) equilibrium are briefly discussed here.

A. The General Model

1. *Analytical Results*

a. Model Specifications

Consider the model

$$x_i(t+1) = \frac{F_i[x_i(t); i = 1, 2, \ldots, I]}{\sum_{j=1}^{I} F_j[x_i(t); i = 1, 2, \ldots, I]}, \qquad \text{(III.A.1.1)}$$

$$F_i > 0, \qquad 0 < x_i < 1; \qquad i = 1, 2, \ldots, I,$$

$$\sum_i x_i(t+1) = 1,$$

$$0 < x_i(0) < 1; \qquad \sum_i x_i(0) = 1.$$

The following equations define the coordinates $x_1^*, x_2^*, \ldots, x_I^*$ of the equilibrium states

$$x_i^* = \frac{1}{1 + \dfrac{1}{F_i^*} \sum_{j=1, j \neq i}^{I} F_j^*}, \qquad \text{(III.A.1.2)}$$

or by utilizing the antiregional notation ($x_i^{-*} = 1 - x_i^*$)

$$\frac{1 - x_i^*}{x_i^*} = \frac{1}{F_i^*} \sum_{\substack{j=1 \\ j \neq i}}^{I} F_j^* = \frac{x_i^{-*}}{x_i^*}; \qquad i = 1, 2, \ldots, I, \qquad \text{(III.A.1.3)}$$

or, stating in a slightly different manner

$$\frac{x_i^{-*} F_i^*}{x_i^*} = \sum_{\substack{j=1 \\ j \neq i}}^{I} F_j^*, \qquad \text{(III.A.1.4)}$$

indicating that the left-hand side of the above is nothing but the sum, in the original community of I regions, of all the other regions' locational advantages.

b. Elasticity of Growth Properties

The following general properties hold

$$\frac{\partial x_i(t+1)}{\partial F_i} = \frac{1}{\sum_j F_j} - \frac{F_i}{(\sum_j F_j)^2} = \frac{x_i(t+1)}{F_i} - \frac{x_i^2(t+1)}{F_i}$$

$$= \frac{x_i(t+1)[1 - x_i(t+1)]}{F_i} = \frac{x_i(t+1)x_i^-(t+1)}{F_i} > 0. \qquad \text{(III.A.1.5)}$$

If we designate by $\varepsilon_i(t+1, t)$ the elasticity of growth in $x_i(t+1)$ with respect to its locational advantage F_i, then

$$\varepsilon_i(t+1,t) = \frac{\partial x_i(t+1)}{\partial F_i}\bigg/\frac{x_i(t+1)}{F_i} \quad \text{and} \quad \varepsilon_i(t+1,t) = x_i^-(t+1), \tag{III.A.1.6}$$

i.e., the elasticity of growth in the portion of the population residing in region i with respect to its locational advantages F_i is equal to its antiregional share (i.e., the portion of the population at all other regions.) The intralocational effect, given by condition (III.A.1.5), is always positive; whereas, the cross-effect (i.e., the interlocational effect) is always negative

$$\frac{\partial x_i(t+1)}{\partial F_j} = \frac{-F_i}{(\sum_j F_j)^2} = -\frac{x_i(t+1)}{\sum_j F_j} = -\frac{x_i(t+1)^2}{F_i}$$

$$= -\frac{x_i(t+1)x_j(t+1)}{F_j} < 0, \tag{III.A.1.7}$$

the last since

$$\frac{x_i(t+1)}{x_j(t+1)} = \frac{F_i}{F_j}. \tag{III.A.1.8}$$

Consider three locations, i, j, k, and let us examine the lagged interlocational effects of one upon the other two; one has immediately

$$\frac{\partial x_i(t+1)}{\partial F_j} = -\frac{x_i(t+1)x_j(t+1)}{F_j}, \tag{III.A.1.9}$$

$$\frac{\partial x_k(t+1)}{\partial F_j} = -\frac{x_k(t+1)x_j(t+1)}{F_j}, \tag{III.A.1.10}$$

and by dividing both sides one obtains

$$\frac{\partial x_i(t+1)}{\partial F_j}\bigg/\frac{\partial x_k(t+1)}{\partial F_j} = \frac{x_i(t+1)}{x_k(t+1)} = \frac{F_i}{F_k}. \tag{III.A.1.11}$$

Thus, current interlocational effects depend upon the magnitude of the current (relative) stock size; these, in turn, are directly proportional to the current locational advantages.

c. Slope Properties

The slope $s_{ii}(t+1,t)$ is given by

$$s_{ii}(t+1,t) = \frac{\partial x_i(t+1)}{\partial x_i(t)} = \frac{\partial x_i(t+1)}{\partial F_i}\cdot\frac{\partial F_i}{\partial x_i}\bigg/F_i - \frac{x_i(t+1)^2}{F_i}\sum_{j\neq i}\frac{\partial F_j}{\partial x_i} \gtrless 0, \tag{III.A.1.12}$$

the latter because of the elasticity form. Further

$$\frac{\partial x_i(t+1)}{\partial x_i(t)} = \frac{x_i(t+1)^2}{1-F_i}\sum_{j\neq i}\frac{\partial F_j}{\partial x_i(t)} \gtrless 0. \tag{III.A.1.13}$$

If $F_i = 1$, i.e., the locational advantages are normalized by the numéraire location's advantages, then

$$\sum_{j \neq i} \frac{\partial F_j}{\partial x_i(t)} = 0, \tag{III.A.1.14}$$

which implies that there is a zero-sum game in comparative advantages as a result of change in any region's population size. A cross-locational effect is given by

$$s_{ij}(t+1, t) = \frac{x_i(t+1)}{F_i}\left[\frac{\partial F_i}{\partial x_j(t)} - x_i(t+1)\sum_h \frac{\partial F_h}{\partial x_j(t)}\right] \gtrless 0. \tag{III.A.1.15}$$

At all time periods, since $\sum_i x_i(t+1) = 1$, the condition holds

$$\sum_i s_{ij}(t+1, t) = \sum_i \frac{\partial x_i(t+1)}{\partial x_j(t)} = 0 \quad \text{for all} \quad j = 1, 2, \ldots, I. \tag{III.A.1.16}$$

In a complete form the interlocational effects are presented by the Jacobi matrix, i.e., the slope-matrix

$$J = \|s_{ij}(t+1, t)\| = \left\|\frac{\partial x_i(t+1)}{\partial x_j(t)}\right\|$$
$$= \left\|\frac{x_i(t+1)}{F_i}\left[\frac{\partial F_i}{\partial x_j(t)} - x_i(t+1)\sum_h \frac{\partial F_h}{\partial x_j(t)}\right]\right\|. \tag{III.A.1.17}$$

Due to the condition (III.A.1.16) the determinant of this matrix—the Jacobian—is equal to zero

$$\det J = 0,$$

i.e., one of the eigenvalues of the Jacobi matrix is equal to zero. At the equilibrium $\bar{x}^* = (x_1^*, x_2^*, \ldots, x_I^*)$ the Jacobi slope-matrix is

$$J^* = \|s_{ij}^*\| = \left\|\frac{x_i^*}{F_i^*}\left(\frac{\partial F_i^*}{\partial x_j^*} - x_i^* \sum_h \frac{\partial F_h^*}{\partial x_j^*}\right)\right\|. \tag{III.A.1.18}$$

As is well known from the von Neumann stability theorem for systems of nonlinear difference equations (see Saaty (1981), p. 168), the equilibrium $\bar{x}^*$ is stable if and only if all eigenvalues λ^* of the Jacobi matrix J^* satisfy the condition

$$|\lambda^*| < 1. \tag{III.A.1.19}$$

For specific cases it is possible to evaluate analytically these eigenvalues and to obtain the geometrical form of the domains of stability of equilibria.

2. *Ranking of Stocks According to Size*

We will attempt next to rank these x_i^*'s according to their equilibrium size

$$x_1^* < x_2^* < \cdots < x_i^* < \cdots < x_I^*. \tag{III.A.2.1}$$

So that if $h < i$, then

$$x_h^* < x_i^* \quad \rightarrow \quad \bar{x}_h^{-*} > \bar{x}_i^{-*}, \tag{III.A.2.2}$$

which implies

$$\frac{1}{1+\dfrac{1}{F_h^*}\sum_{j\neq h}^{I} F_j^*} < \frac{1}{1+\dfrac{1}{F_i^*}\sum_{j\neq i}^{I} F_j^*}, \qquad \text{(III.A.2.3)}$$

or put differently

$$\frac{1}{F_h^*}\sum_{\substack{j=1\\ j\neq h}}^{I} F_j^* > \frac{1}{F_i^*}\sum_{\substack{j=1\\ j\neq i}}^{I} F_j^* \quad \rightarrow \quad \frac{x_h^{-*}}{x_h^*} > \frac{x_i^{-*}}{x_i^*}, \qquad \text{(III.A.2.4)}$$

which finally implies

$$\frac{F_i^*}{F_h^*} > \frac{\sum_{j\neq i} F_j^*}{\sum_{j\neq h} F_j^*}. \qquad \text{(III.A.2.5)}$$

If

$$\sum_{s=1} F_s^* = F^* \qquad \text{then} \qquad \frac{F_i^*}{F_h^*} > \frac{F^* - F_i^*}{F^* - F_h^*} = \frac{F_i^{-*}}{F_h^{-*}}.$$

Thus, at equilibrium for a location i's stock to exceed that of a competing location h, the locational advantages of i at equilibrium (F_i^*) must exceed that of h (F_h^*) weighed by their antiregional share (F_i^{-*}/F_h^{-*}).

Consider that the F_i functions move within certain limits

$$0 < k_i \le F_i[x_1(t), x_2(t), \ldots, x_I(t)] \le K_i < \infty, \qquad \text{(III.A.2.6)}$$

where the parameters k_i and K_i, $i = 1, 2, \ldots, I$, do not depend on t. Each location's population abundance $x_i(t+1)$ moves within these bounded strips. They are directly obtainable from the condition

$$\frac{x_i(t+1)}{x_j(t+1)} = \frac{F_i}{F_j},$$

which provides

$$\frac{k_i}{K_j} \le \frac{x_i(t+1)}{x_j(t+1)} \le \frac{K_i}{k_j}. \qquad \text{(III.A.2.7)}$$

The above implies, in turn, that

$$\frac{1}{K_j}\sum_{i\neq j} k_i \le \sum_{i\neq j}\frac{x_i(t+1)}{x_j(t+1)} = \frac{1-x_j(t+1)}{x_j(t+1)}$$

$$\le \frac{1}{k_j}\sum_{i\neq j} K_i,$$

and therefore

$$\frac{1}{1+\dfrac{1}{k_j}\sum_{i\neq j} K_j} \le x_j(t+1) \le \frac{1}{1+\dfrac{1}{K_j}\sum_{i\neq j} k_j}, \qquad j = 1, 2, \ldots, I. \qquad \text{(III.A.2.8)}$$

3. *Trajectory Domains for the One-Stock, Three-Location Model*

a. The Jacobi Matrix of the System

To define the area where trajectories of the one-stock, three-location model move, we need to look at the Jacobi slope-matrix of the dynamic system. At the equilibrium the Jacobi slope-matrix J^* for the one-stock, three-location model has the form

$$J^* = \|s_{ij}^*\| = \left\| \frac{x_i^*}{F_i^*}\left[\frac{\partial F_i^*}{\partial x_j^*} - x_i^*\left(\frac{\partial F_1^*}{\partial x_j^*} + \frac{\partial F_2^*}{\partial x_j^*} + \frac{\partial F_3^*}{\partial x_j^*}\right)\right] \right\|, \qquad i, j = 1, \ldots, 3. \tag{III.A.3.1}$$

The characteristic equation for this matrix is

$$\lambda^3 - \mathrm{Tr}\, J^* \lambda^2 + \Delta\lambda + \det J^* = 0, \tag{III.A.3.2}$$

where

$$\det J^* = \begin{vmatrix} s_{11}^* & s_{12}^* & s_{13}^* \\ s_{21}^* & s_{22}^* & s_{23}^* \\ s_{31}^* & s_{32}^* & s_{33}^* \end{vmatrix} = 0,$$

$$\mathrm{Tr}\, J^* = \sum_{i=1}^{3} s_{ii}^* = \frac{x_1^*}{F_1^*}\left[\left(\frac{\partial F_1^*}{\partial x_1^*} + \frac{\partial F_2^*}{\partial x_2^*} + \frac{\partial F_3^*}{\partial x_3^*}\right) - \sum_{i=1}^{3} x_i^* \left(\frac{\partial F_1^*}{\partial x_i^*} + \frac{\partial F_2^*}{\partial x_i^*} + \frac{\partial F_3^*}{\partial x_i^*}\right)\right], \tag{III.A.3.3}$$

and Δ is the sum of the main second-order determinants of J^*,

$$\Delta = \Delta_1 + \Delta_2 + \Delta_3. \tag{III.A.3.4}$$

Second-order determinants are supplied in Appendix I.

Since $\det J^* = 0$, then nonzero eigenvalues of the Jacobi matrix J^* are found in the solution of the quadratic equation

$$\lambda^2 - \mathrm{Tr}\, J^* \lambda + \Delta = 0. \tag{III.A.3.5}$$

These solutions are

$$\lambda_{1,2} = \tfrac{1}{2}[\mathrm{Tr}\, J^* \pm \sqrt{(\mathrm{Tr}\, J^*)^2 - 4\Delta}]. \tag{III.A.3.6}$$

If

$$(\mathrm{Tr}\, J^*)^2 - 4\Delta < 0, \tag{III.A.3.7}$$

then the eigenvalues are the complex conjugate numbers

$$\lambda_{1,2} = a \pm ib = \tfrac{1}{2}[\mathrm{Tr}\, J^* \pm i\sqrt{4\Delta - (\mathrm{Tr}\, J^*)^2}\,]. \tag{III.A.3.8}$$

Therefore, the condition $|\lambda_{1,2}| < 1$ means that

$$a^2 + b^2 < 1,$$

or

$$a^2 + b^2 = \tfrac{1}{4}[(\text{Tr } J^*)^2 + 4\Delta - (\text{Tr } J^*)^2] = \Delta < 1. \qquad \text{(III.A.3.9)}$$

b. Domains of Stability

Conditions (III.A.3.7, 8) give the following inequalities, defining the domain of stability of the dynamic equilibria,

$$\tfrac{1}{4}(\text{Tr } J^*)^2 < \Delta < 1. \qquad \text{(III.A.3.10)}$$

If

$$(\text{Tr } J^*)^2 - 4\Delta \geq 0, \qquad \text{(III.A.3.11)}$$

then the characteristic equation (III.A.3.5) has two real solutions. The condition (III.A.1.19) of stability of equilibria means (see Figure 15) that the parabola

$$y = \lambda^2 - \text{Tr } J^* \lambda + \Delta$$

intersects the λ-axis within the interval $(-1, 1)$.

This is possible if and only if the parabola obtains the minimal value at the point $\frac{1}{2}$ Tr J^* within the interval $(-1, 1)$

$$-1 < \tfrac{1}{2} \text{Tr } J^* < 1, \qquad \text{(III.A.3.12)}$$

and the values $y(-1)$, $y(+1)$ are positive

$$y(-1) = 1 + \text{Tr } J^* + \Delta > 0, \qquad \text{(III.A.3.13)}$$

$$y(+1) = 1 - \text{Tr } J^* + \Delta > 0, \qquad \text{(III.A.3.14)}$$

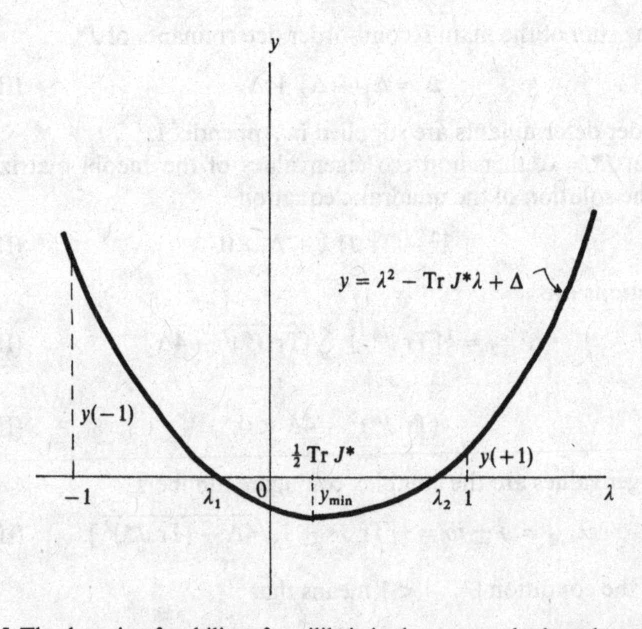

FIGURE 15. The domain of stability of equilibria in the one-stock, three-location model.

Conditions (III.A.3.10–14) give the following inequalities

$$\Delta \leq \tfrac{1}{4}(\mathrm{Tr}\, J^*)^2,$$

$$-2 \leq \mathrm{Tr}\, J^* \leq 2, \qquad \text{(III.A.3.15)}$$

$$\Delta + 1 \pm \mathrm{Tr}\, J^* > 0.$$

Conditions (III.A.3.9) and (III.A.3.15) describe the domain of stability of equilibria for the one-stock, three-location case. These conditions can be summarized in the following way

$$-1 \pm \mathrm{Tr}\, J^* < \Delta < 1. \qquad \text{(III.A.3.16)}$$

Moreover, the condition $-1 + \mathrm{Tr}\, J^* = \Delta$ means that the biggest real eigenvalue is $\lambda_2 = 1 > \lambda_1 \geq -1$; condition $-1 - \mathrm{Tr}\, J^* = \Delta$ means, further, that the smallest real eigenvalue is $\lambda_1 = -1 < \lambda_2 \leq 1$; and finally condition $\Delta = 1$ implies that the two complex eigenvalues λ_1, λ_2 have the same absolute value $|\lambda_1| = |\lambda_2| = 1$.

Geometrically, the domain of stability of equilibria is a triangle ABC in the space of parameters Δ, $\mathrm{Tr}\, J^*$ (see Figure 16). The vertices of this triangle have the following coordinates

$$A: \begin{cases} \Delta = 1, \\ \mathrm{Tr}\, J^* = -2, \end{cases} \quad B: \begin{cases} \Delta = 1, \\ \mathrm{Tr}\, J^* = 2, \end{cases} \quad C: \begin{cases} \Delta = -1, \\ \mathrm{Tr}\, J^* = 0. \end{cases} \qquad \text{(III.A.3.17)}$$

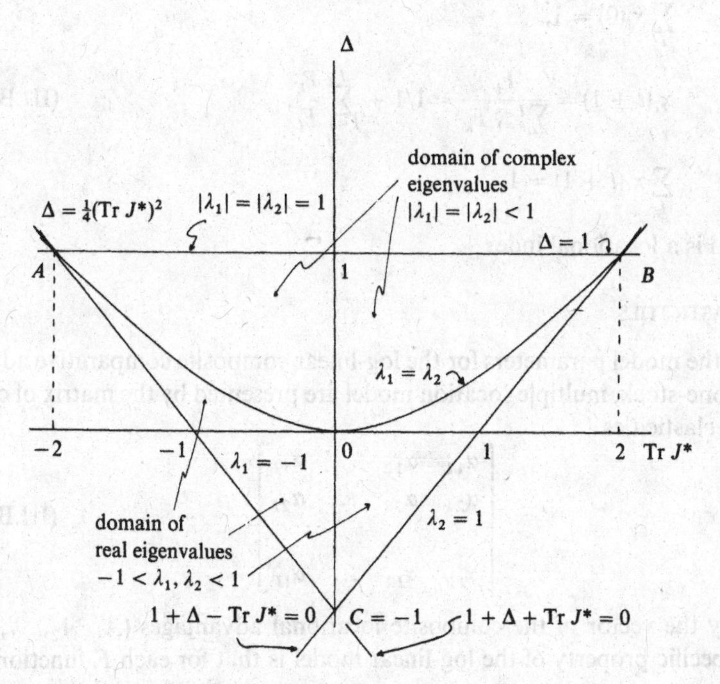

FIGURE 16. Domains of stability of equilibria in the space of parameters (Δ, $\mathrm{Tr}\, J^*$).

The side AB of this triangle presents the complex eigenvalues λ_1, λ_2, so that $|\lambda_1| = |\lambda_2| = 1$; BC corresponds to the case where $\lambda_2 = 1 > \lambda_1 > -1$, and AC to the case where $-1 = \lambda_1 < \lambda_2 < 1$.

The conditions for stability (III.A.3.10) are the outcome of the general rule that the polynomial $x^2 + a_1 x + a_2$ has roots less than one in absolute value, if and only if $-1 \pm a_1 < a_2 < 1$. In the general case of the polynomial $x^n + a_1 x^{n-1} + a_2 x^{n-2} + \cdots + a_n$ the Routh–Hurwitz condition for all roots to be less than unity in absolute value can be found in Samuelson (1941).

B. The Log-Linear Comparative Advantages Model

1. *Analytical Properties*

a. Specifications

In this section we consider the following specification of the F_i function:

$$F_i = A_i \sum_{j=1}^{I} x_j(t)^{\alpha_{ij}}, \tag{III.B.1.1}$$

$$-\infty < \alpha_{ij} < +\infty; \qquad i, j = 1, 2, \ldots, I,$$

$$F_i > 0; \qquad A_i > 0; \qquad 0 < x_i(0) < 1; \qquad i = 1, 2, \ldots, I,$$

$$\sum_i x_i(0) = 1,$$

$$x_i(t+1) = \frac{F_i}{\sum_{k=1}^{I} F_k} = 1/1 + \sum_{\substack{k=1 \\ k \neq i}}^{I} \frac{F_k}{F_i}, \tag{III.B.1.2}$$

$$\sum_i x_i(t+1) = 1,$$

where i is a locational index.

b. Elasticities

Thus, the model parameters for the log-linear composite comparative advantages one-stock, multiple-location model are presented by the matrix of composite elasticities

$$\begin{bmatrix} \alpha_{11} & \alpha_{12} & \ldots & \alpha_{1I} \\ \alpha_{21} & \alpha_{22} & \ldots & \alpha_{2I} \\ & & \vdots & \\ \alpha_{I1} & \alpha_{I2} & \ldots & \alpha_{II} \end{bmatrix}, \tag{III.B.1.3}$$

and by the vector of the composite locational advantages $(A_1, A_2, \ldots, A_I)$. The specific property of the log-linear model is that for each F_i function the following holds

$$\frac{\partial F_i}{\partial x_j(t)} = \frac{\alpha_{ij}}{x_j(t)} F_i, \qquad i, j = 1, 2, \ldots, J. \tag{III.B.1.4}$$

c. Slopes

The components (slopes) of the Jacobi slope-matrix will be (see (III.A.1.15) and (III.A.1.8))

$$\begin{aligned}\frac{\partial x_i(t+1)}{\partial x_j(t)} &= \frac{x_i(t+1)}{F_i}\left[\frac{\alpha_{ij}}{x_j(t)} F_i - x_i(t+1) \sum_h \frac{\alpha_{hj}}{x_j(t)} F_h\right] \\ &= \frac{x_i(t+1)}{x_j(t)}\left[\alpha_{ij} - x_i(t+1) \sum_h \alpha_{hj} \frac{F_h}{F_i}\right] \\ &= \frac{x_i(t+1)}{x_j(t)}\left[\alpha_{ij} - \sum_h \alpha_{hj} x_h(t+1)\right].\end{aligned} \tag{III.B.1.5}$$

At the equilibrium the Jacobi slope-matrix $J^* = \|s_{ij}^*\|$ will include slopes which are rational functions of the parameters $x_1^*, x_2^*, \ldots, x_I^*$ of the equilibrium state

$$s_{ij}^* = \frac{x_i^*}{x_j^*}\left[\alpha_{ij} - \sum_h \alpha_{hj} x_h^*\right], \qquad i, j = 1, 2, \ldots, I. \tag{III.B.1.6}$$

d. Stability Analysis

According to the von Neumann theorem, the equilibrium state $x_1^*, \ldots, x_I^*$ will be stable, if for each eigenvalue λ^* of the Jacobi matrix J^* the following condition holds

$$|\lambda^*| < 1. \tag{III.B.1.7}$$

The composite comparative advantages function G_i is

$$G_i = \sum_{\substack{j=1 \\ j \neq i}}^{I} \frac{A_j}{A_i} \prod_{k=1}^{I} x_k(t)^{\beta_{kji}}, \tag{III.B.1.8}$$

where $\beta_{kij} = \alpha_{kj} - \alpha_{ki}$. Relative to a numéraire (place of reference) location I, each location's population abundance is

$$\begin{aligned}\frac{x_i(t+1)}{x_I(t+1)} &= \frac{A_i}{A_I} \prod_{k=1}^{I} x_k(t)^{\alpha_{ik} - \alpha_{Ik}} \\ &= \frac{A_i}{A_I} \prod_{k=1}^{I-1} \left[\frac{x_k(t)}{x_I(t)}\right]^{\alpha_{ik} - \alpha_{Ik}} x_I(t)^{\sum_{k=1}^{I} (\alpha_{ik} - \alpha_{Ik})},\end{aligned} \tag{III.B.1.9}$$

and

$$\sum_i^{I-1} \frac{x_i(t+1)}{x_I(t+1)} < 1.$$

In general, these specifications made it impossible to track analytically the properties of stability of the system. We need some simplification to gauge the system's behavior in certain points in the parameter space.

e. A Special Case

A special case of mathematical interest is the following

$$\sum_{k=1}^{I} (\alpha_{ik} - \alpha_{Ik}) = 0. \tag{III.B.1.10}$$

If we denote by

$$y_i(t+1) = \frac{x_i(t+1)}{x_I(t+1)}; \qquad \sum_i y_i(t+1) < 1; \qquad y_i(t+1) \geq 0;$$

$$B_i = \frac{A_i}{A_I}; \qquad \beta_{ik} = \alpha_{ik} - \alpha_{Ik},$$

then

$$y_i(t+1) = B_i \prod_{k=1}^{I-1} y_k(t)^{\beta_{ik}}, \tag{III.B.1.11}$$

and

$$\ln y_i(t+1) = \ln B_i + \prod_{k=1}^{I-1} \beta_{ik} \ln y_k(t). \tag{III.B.1.12}$$

If we denote by $z_i = \ln y_i$ and $C_i = \ln B_i$, then

$$z_i(t+1) = C_i + \prod_{k=1}^{I-1} \beta_{ik} z_k(t), \tag{III.B.1.13}$$

which at equilibrium provides

$$z_i^* = C_i + \prod_{k=1}^{I-1} \beta_{ik} z_k^*. \tag{III.B.1.14}$$

This is a system of $(I-1)$ equations on $(I-1)$ unknowns of the form

$$Mz^* = C, \tag{III.B.1.15}$$

where the matrix M is given by

$$M = \begin{bmatrix} 1-\beta_{1,1} & -\beta_{1,2} & \cdots & -\beta_{1,I-1} \\ -\beta_{2,1} & 1-\beta_{2,2} & \cdots & -\beta_{2,I-1} \\ \vdots & \vdots & & \vdots \\ -\beta_{I-1,1} & -\beta_{I-1,2} & \cdots & 1-\beta_{I-1,I-1} \end{bmatrix}.$$

The necessary condition for this special case to have a unique admissible equilibrium is that its determinant be nonzero.

It is very unlikely that any arbitrary (β_{ij}) matrix, although it may have $\det M \neq 0$, will also satisfy the conditions $y_i \geq 0$; thus, it is not likely that the very special case indicated by condition (III.B.1.10) will have a stable, unique, admissible equilibrium. Consequently, it is even more unlikely that the more general system (III.B.1.1) will do.

f. Some General Statements for the Multiple-Location Model

Turning back to the original specifications, at equilibrium one obtains

$$x_i^* = 1/1 + \sum_{\substack{k=1 \\ k \neq i}}^{I} \frac{F_k^*(x_k^*; k = 1, 2, \ldots, I)}{F_i^*(x_i^*; i = 1, 2, \ldots, I)}$$

$$= 1/1 + \sum_{\substack{k=1 \\ k \neq i}}^{I} \frac{A_k \sum_{j=1}^{I} x_j^{*\alpha_{kj}}}{\prod_{j=1}^{I} x_j^{*\alpha_{ij}}}$$

$$= 1/1 + \frac{1}{A_i} \sum_{\substack{k=1 \\ k \neq i}}^{I} A_k \prod_{j=1}^{I} x_j^{*\alpha_{kj}-\alpha_{ij}}, \qquad \text{(III.B.1.16)}$$

or

$$\frac{1 - x_i^*}{x_i^*} = \frac{1}{A_i} \sum_{\substack{k=1 \\ k \neq i}}^{I} A_k \prod_{j=1}^{I} x_j^{*\alpha_{kj}-\alpha_{ij}}, \qquad \text{(III.B.1.17)}$$

which can also be written as

$$A_i = \sum_{\substack{k=1 \\ k \neq i}}^{I} A_k \frac{x_i^*}{1 - x_i^*} \prod_{j=1}^{I} x_j^{*\alpha_{kj}-\alpha_{ij}}; \quad i = 1, 2, \ldots, I. \qquad \text{(III.B.1.18)}$$

This is the multidimensional Φ function, equivalent to (II.A.2.29) for the two-dimensional case. The Φ' functions are

$$\frac{\partial \Phi_i}{\partial x_h^*} = \sum_{\substack{k=1 \\ k \neq i}}^{I} A_k \frac{x_i^*}{1 - x_i^*} \frac{\partial}{\partial x_h^*} \prod_{j=1}^{I} x_j^{*\alpha_{kj}-\alpha_{ij}}; \quad h = 1, 2, \ldots, I. \qquad \text{(III.B.1.19)}$$

2. *Numerical Analysis for the One-Stock, Multiple-Location Model*

In this subsection the focus is on the one-stock, I-location model. Subsection (a) contains the case where $I = 3$, subsection (b) where $I = 4$, and subsection (c) where $I = 5$, as well as certain results where $I = 10$. Empirical evidence using data on the relative population abundance of the U.S. regions, looked at from a three- and four-region spatial disaggregation, is also presented.

A number of innovative features are shown to be contained in the universal relative discrete dynamics algorithm. These results include the onset of a limited period-doubling phase, that could be labeled "local" and "partial" turbulence; strange, pure, and hybrid attractors and containers found in the deterministic motion recorded in certain regions of chaos in the parameter space of the algorithm. For socio-spatial systems, the chaotic regimes may be more interesting than the regions of period-doubling (or any other) periodic motion. These must not be the only new phenomena hidden in this discrete dynamics algorithm and future research will very likely reveal more.

At first, the effects of increasing the model's dimensionality upon the analytical properties and the computational capacity (no matter the machine used for computing) are reviewed briefly. Given any arbitrary step 10^{ξ} on the bifurcation parameters $\underline{A}$, and any arbitrary step χ on the exponents α with bounds

$$10^{-\lambda} \le A_i \le 10^{\lambda}, \qquad i = 2, 3, \ldots, I, \quad A_1 = 1;$$

$$-\mu \le \alpha_{ij} \le \mu, \qquad i = 2, 3, \ldots, I, \quad j = 1, 2, \ldots, I, \quad \alpha_{1j} = 0,$$

and orbits, i.e., set of iterations, of magnitude T, the total possible number of computations is

$$e = \left(\frac{2\lambda}{\xi} + 1\right)^{(I-1)} T\left(\frac{2\mu}{\chi} + 1\right)^{(I-1)I},$$

although the actual number depends on the stopping rule used to identify whether or not the steady state has been reached.

Clearly, as the dimensionality of the problem increases, the number of computations per unit of computer time (again, regardless of the machine used) decreases exponentially as the square of the problem's dimension I. Due to the fact that the number of iterations increase in the A's as the power $(I - 1)$, whereas the computations increase in the exponents approximately as I^2, greater ranges can be examined in the bifurcation parameters, and in more fine spatial detail, than in the exponents, i.e., the structural parameters.

A second issue facing the analyst is associated with the computational accuracy of the iterative process as the orbits become longer, i.e., T increases. There is a point beyond which the results obtained are the outcome of random disturbances or noise, due to computer approximation (a level which depends on the machine used).

A third issue is related to the machine over- (under-) flow which affects the behavior of the iterative process in the case of competitive exclusion equilibria. The over- (under-) flow is obtained when the fixed-point algorithm, in very few steps, converges to values in the state variables $\underline{x}$ close to $10^{\pm k}$, where k depends on the machine used. For a VAX-11/750 system over- (under-) flow is obtained when $k = \pm 38$. Using this system with $\lambda = 10$, and a stopping rule for having found a fixed point at 10^{-8}, experiments suggest the following: in the three-dimensional case, over-flow occurs when all exponents are equal to $\bar{\alpha}$ and varies in the range $-2.5 < \bar{\alpha} < -2.0$. For the four-dimensional and identical exponents case the floor drops to $-2.0 < \bar{\alpha} < -1.5$; in the five-dimensional model the floor is $-1.5 < \bar{\alpha} < -1.0$. A CYBER-205 system extends these limits to $-3.0 < \bar{\alpha} < -2.5$ for the five-dimensional case.

Although these results are applicable only for the VAX system and under the conditions stated, the point is clear regarding all computational specifications and/or the system used: as the problem's dimensionality increases, smaller ranges in the parameters' subspaces can be looked at and increasingly stable (in the state variables) regions are possible to examine.

One might obtain a glimpse in these findings of a picture connecting complexity and stability; in relative dynamics their links could be slightly different than in May's original argument (1974). Here, the area of turbulence is squeezed down to extinction as I increases (from a computational standpoint) and the range of α where the system produces over- (under-) flow (competitive exclusion) increases.

There is some dimension I^* where only stable (fixed-point) equilibria can be obtained within the allowable region (where no over- or underflow occurs). As I increases, there is another threshold dimension I^{**} reached where no computation is feasible (i.e., only competitive exclusion prevails). It must be kept in mind that these thresholds vary depending on what machine is used. Thus, the state of the discrete dynamic system is tied to the hardware of the computing system, rather than the software, something which could be potentially quite informative for socio-spatial dynamics and its epistemological foundations.

a. Three Locations

In this case, the model specifications take the form, from (III.B.1.1, 2),

$$F_i = A_i \prod_{j=1}^{3} x_j(t)^{\alpha_{ij}}, \qquad i = 1, 2, 3, \tag{III.B.2.a.1}$$

$$-\infty \le \alpha_{ij} \le +\infty, \qquad i, j = 1, 2, 3,$$

$$F_i > 0, A_i > 0, \qquad 0 \le x_i(0) \le 1, \quad i = 1, 2, 3,$$

so that

$$0 < x_i(t) < 1, \qquad i = 1, 2, 3; \quad t = 1, 2, \ldots, T,$$

$$\sum_i x_i(t) = 1, \qquad t = 1, 2, \ldots, T, \tag{III.B.2.a.2}$$

$$x_i(t+1) = \frac{F_i}{\sum_{j=1}^{3} F_j} = 1/1 + \frac{\sum_{j=1, j\neq i}^{3} F_j}{F_i}, \qquad i = 1, 2, 3. \tag{III.B.2.a.3}$$

From the locational advantages functions, F_i, the comparative advantages, G_i, can be computed

$$G_i = \sum_{\substack{j=1 \\ j\neq i}}^{3} \frac{F_j}{F_i}, \qquad i = 1, 2, 3, \tag{III.B.2.a.4}$$

which, under specifications (III.B.2.a.1), produce

$$G_i = \sum_{\substack{j=1 \\ j\neq i}}^{3} \frac{A_j}{A_i} \prod_{k=1}^{3} x_k(t)^{\beta_{kji}}, \qquad i = 1, 2, 3, \tag{III.B.2.a.5}$$

where

$$\beta_{kji} = \alpha_{jk} - \alpha_{ij}, \qquad k, j, i = 1, 2, 3, \tag{III.B.2.a.6}$$

so that condition (III.B.2.a.3) becomes

$$x_i(t+1) = 1/1 + G_i. \qquad \text{(III.B.2.a.7)}$$

In the three-location, one-stock case the above condition gives the following result

$$2 + \sum_{i=1}^{3} G_i = \prod_{i=1}^{3} G_i. \qquad \text{(III.B.2.a.8)}$$

In the two-location, one-stock case the comparative advantages are related as follows

$$1 = \prod_{i=1}^{2} G_i, \qquad \text{(III.B.2.a.9)}$$

whereas, in the four-dimensional case, the condition holds

$$3 + \sum_{i=1}^{4} G_i + \sum_{i=1}^{4} \sum_{\substack{j=i \\ j\neq 1}}^{4} G_i G_j = \prod_{i=1}^{4} G_i. \qquad \text{(III.B.2.a.10)}$$

From the above analysis one can easily deduce the following relationship for the I-dimensional case

$$(I-1) + \sum_{i=1}^{I} G_i + \sum_{i=1}^{I} \sum_{\substack{j=i \\ j\neq 1}}^{I} G_i G_j + \cdots + \sum_{i=1}^{I} \sum_{\substack{j=i \\ j\neq 1}}^{I} \cdots \sum_{\substack{k\neq i \\ k\neq j \\ \vdots \\ k=1}}^{I} G_i G_j \cdots G_k = \sum_{i=1}^{I} G_i. \qquad \text{(III.B.2.a.11)}$$

Setting region one as the numéraire location, we can set $A_1 = 1$, $\alpha_{1j} = 0$, $j = 1, 2, 3$. Thus, in general, the dimensions in the parameter space of the I-dimensional problem are $(I-1)I$, in the subspace, for a total of $(I^2 - 1)$. For any set of α's and by varying the bifurcation parameters A_2, A_3, one can obtain the behavior of the three state variables in this one-stock, three-location version of the universal relative dynamics model.

i. Graphical Presentation of Domains of Stability: the Möbius Barycentric Coordinates

For the one-stock, three-population, log-linear model the Jacobi slope-matrix J^* gives the following entries Tr J^* and Δ for the characteristic equations (III.A.3.5)

$$\begin{aligned} \operatorname{Tr} J^* = \sum_{i=1}^{3} s_{11}^* &= (\alpha_{11} + \alpha_{22} + \alpha_{23}) - \sqrt{\sum_{h=1}^{3} \sum_{i=1}^{3} \alpha_{hi} x_h^*} \\ &= (\alpha_{11} + \alpha_{22} + \alpha_{23}) - x_1^*(\alpha_{11} + \alpha_{12} + \alpha_{13}) \\ &\quad - x_2^*(\alpha_{21} + \alpha_{22} + \alpha_{23}) - x_3^*(\alpha_{31} + \alpha_{31} + \alpha_{33}), \end{aligned} \qquad \text{(III.B.2.a.i.1)}$$

and (see Appendix II)

$$\Delta = \begin{vmatrix} s_{11}^* & s_{12}^* \\ s_{21}^* & s_{22}^* \end{vmatrix} + \begin{vmatrix} s_{11}^* & s_{13}^* \\ s_{31}^* & s_{32}^* \end{vmatrix} + \begin{vmatrix} s_{22}^* & s_{23}^* \\ s_{32}^* & s_{33}^* \end{vmatrix}$$

$$= \begin{vmatrix} \alpha_{11} & \alpha_{12} & \alpha_{13} \\ \alpha_{21} & \alpha_{22} & \alpha_{23} \\ x_1^* & x_2^* & x_3^* \end{vmatrix} + \begin{vmatrix} \alpha_{11} & \alpha_{12} & \alpha_{13} \\ x_1^* & x_2^* & x_3^* \\ \alpha_{31} & \alpha_{32} & \alpha_{33} \end{vmatrix} + \begin{vmatrix} x_1^* & x_2^* & x_3^* \\ \alpha_{21} & \alpha_{22} & \alpha_{23} \\ \alpha_{31} & \alpha_{32} & \alpha_{33} \end{vmatrix}. \qquad \text{(III.B.2.a.i.2)}$$

For the special choice where $A_1 = 1$ and the matrix of elasticities

$$\begin{bmatrix} 0 & 0 & 0 \\ \alpha_{21} & \alpha_{22} & \alpha_{23} \\ \alpha_{31} & \alpha_{32} & \alpha_{33} \end{bmatrix} \qquad \text{(III.B.2.a.i.3)}$$

(so that the first location is the numéraire location), we have

$$\frac{x_i(t+1)}{x_1(t+1)} = A_i[x_1(t)]^{\alpha_{i1}}[x_2(t)]^{\alpha_{i2}}[x_3(t)]^{\alpha_{i3}}, \qquad i = 2, 3, \qquad \text{(III.B.2.a.i.4)}$$

and for the equilibrium state x_1^*, x_2^*, x_3^*, we obtain

$$\frac{x_i^*}{x_1^*} = A_i(x_1^*)^{\alpha_{i1}}(x_2^*)^{\alpha_{i2}}(x_3^*)^{\alpha_{i3}}, \qquad i = 2, 3. \qquad \text{(III.B.2.a.i.5)}$$

Then the trace is

$$\operatorname{Tr} J^* = (\alpha_{22} + \alpha_{23}) - x_2^*(\alpha_{21} + \alpha_{22} + \alpha_{23}) - x_3^*(\alpha_{31} + \alpha_{32} + \alpha_{33})$$

and the determinant

$$\Delta = x_1^* \begin{vmatrix} \alpha_{22} & \alpha_{23} \\ \alpha_{32} & \alpha_{33} \end{vmatrix} - x_2^* \begin{vmatrix} \alpha_{21} & \alpha_{23} \\ \alpha_{31} & \alpha_{33} \end{vmatrix} + x_3^* \begin{vmatrix} \alpha_{21} & \alpha_{22} \\ \alpha_{31} & \alpha_{32} \end{vmatrix}$$

$$= \begin{vmatrix} x_1^* & x_2^* & x_3^* \\ \alpha_{21} & \alpha_{22} & \alpha_{23} \\ \alpha_{31} & \alpha_{32} & \alpha_{33} \end{vmatrix}. \qquad \text{(III.B.2.a.i.6)}$$

To derive the geometrical description of the domains of stability of equilibria for the one-stock, three-location model we need to introduce the space of the Möbius barycentric coordinates. The scale element of this space is the Möbius equilateral triangle with the unit scale on its sides (see Figure 17). It is possible to measure the barycentric coordinates, x_1, x_2, x_3, of each point $\bar{x}$ in a plane by projecting this point onto the sides of the triangle parallel to the sides.

If the point $\bar{x}$ lies within the Möbius triangle, then its barycentric coordinates, x_1, x_2, x_3, must be positive and the conservation condition holds

$$x_1 + x_2 + x_3 = 1, \qquad 0 \leq x_1, x_2, x_3 \leq 1.$$

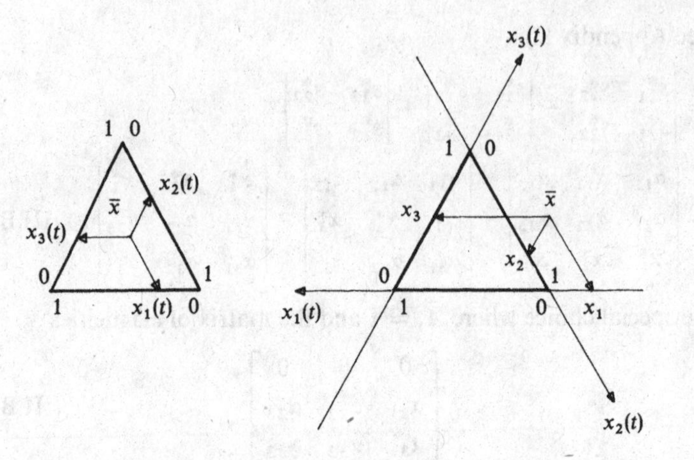

FIGURE 17. Barycentric coordinates in a plane.

If point $\bar{x}$ lies outside the Möbius triangle, then one or both of the barycentric coordinates must be negative or greater than 1, but the condition $x_1 + x_2 + x_3 = 1$ always holds.

Actually, the barycentric coordinates of the point $\bar{x}$ are the coordinates of the center of gravity: if we hang weights, x_1, x_2, x_3, on the vertices of the Möbius triangle, then the triangle's center of gravity will coincide with point $\bar{x}$. The barycentric coordinates are useful for the presentation of trajectories of the one-stock, three-location, relative dynamics model, because in the Möbius triangle all events of all possible dynamic states of the model occur.

To obtain a description of the equilibria stability domain we construct (in the space of barycentric coordinates) the image of this domain, which is the triangle ABC in the space of parameters Δ, Tr J^* (Figure 15). Because the parameters Δ and Tr J^* are linear functions of the coordinates, x_1^*, x_2^*, x_j^*, the image of the triangle ABC is also a triangle. Therefore, we will use the same notation ABC.

The vertices of this image correspond to the solutions of three systems of linear equations

$$A: \begin{cases} \Delta = 1, \\ \operatorname{Tr} J^* = -2, \\ x_1^* + x_2^* + x_3^* = 1, \end{cases} \qquad B: \begin{cases} \Delta = 1, \\ \operatorname{Tr} J^* = 2, \\ x_1^* + x_2^* + x_3^* = 1, \end{cases} \qquad C: \begin{cases} \Delta = -1, \\ \operatorname{Tr} J^* = 0, \\ x_1^* + x_2^* + x_3^* = 1. \end{cases}$$

The domain of stability of equilibria is given by the intersection of the triangle ABC with the basic Möbius triangle. As an example, Figure 18 presents the domain of equilibria for the log-linear model with $A_1 = 1$ and with the matrix of elasticities

$$\begin{bmatrix} 0 & 0 & 0 \\ 1 & 1 & -1 \\ -1 & 1 & 1 \end{bmatrix}.$$

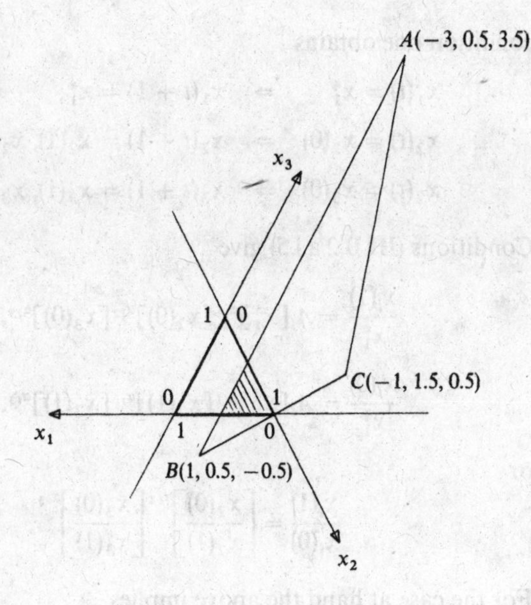

FIGURE 18. A special case. Domain of stability of equilibria for the model $\begin{bmatrix} 0 & 0 & 0 \\ 1 & 1 & -1 \\ -1 & 1 & 1 \end{bmatrix}$ as an intersection of two triangles.

For this model

$$\operatorname{Tr} J^* = 2 - x_2^* - x_3^* = 1 + x_1^*,$$
$$\Delta = 2x_1 + 2x_3 = 2(1 - x_2^*).$$

Coordinates of the points A, B, C are defined by the following systems of equations

$$A: \begin{cases} 2(1 - x_2^*) = 1 \\ 1 + x_1^* = -2 \end{cases} \quad \Rightarrow \quad x_1^* = -3, x_2^* = 0.5, x_3^* = 3.5,$$

$$B: \begin{cases} 3(1 - x_2^*) = 1 \\ 1 + x_1^* = 2 \end{cases} \quad \Rightarrow \quad x_1^* = 1, x_2^* = 0.5, x_3^* = -0.5,$$

$$C: \begin{cases} 2(1 - x_2^*) = -1 \\ 1 + x_1^* = 0 \end{cases} \quad \Rightarrow \quad x_1^* = -1, x_2^* = 1.5, x_3^* = 0.5.$$

so that the domain of stability of equilibria is the shaded area shown in Figure 18.

ii. Local and Partial Turbulence

This phenomenon is associated with the following onset of periodic motion in the three-location, one-stock problem from a stable fixed point: one location's stock, x_1, exhibits a fixed-point equilibrium x_1^*, whereas the other two locations' stocks (x_2, x_3) demonstrate a stable two-period cycle. Under this condition one obtains

$$\begin{aligned} x_1(t) &= x_1^* &&\Rightarrow\quad x_1(t+1) = x_1^*, \\ x_2(t) &= x_2(0) &&\Rightarrow\quad x_2(t+1) = x_2(1),\ x_2(0) \neq x_2(1), \\ x_3(t) &= x_3(0) &&\Rightarrow\quad x_3(t+1) = x_3(1),\ x_3(0) \neq x_3(1). \end{aligned} \tag{III.B.2.a.ii.1}$$

Conditions (III.B.2.a.i.5) give

$$\begin{aligned} \frac{x_i(1)}{x_1^*} &= A_i[x_1^*]^{\alpha_{i1}}[x_2(0)]^{\alpha_{i2}}[x_3(0)]^{\alpha_{i3}}, \qquad i = 2, 3, \\ \frac{x_i(0)}{x_1^*} &= A_i[x_1^*]^{\alpha_{i1}}[x_2(1)]^{\alpha_{i2}}[x_3(1)]^{\alpha_{i3}}, \end{aligned} \tag{III.B.2.a.ii.2}$$

or

$$\frac{x_i(1)}{x_i(0)} = \left[\frac{x_2(0)}{x_2(1)}\right]^{\alpha_{i2}} \left[\frac{x_3(0)}{x_3(1)}\right]^{\alpha_{i3}}, \qquad i = 2, 3. \tag{III.B.2.a.ii.3}$$

For the case at hand the above implies

$$\left[\frac{x_2(0)}{x_2(1)}\right]^{\alpha_{22}+1} \left[\frac{x_3(0)}{x_3(1)}\right]^{\alpha_{23}} = \left[\frac{x_2(0)}{x_2(1)}\right]^{\alpha_{32}} \left[\frac{x_3(0)}{x_3(1)}\right]^{\alpha_{33}+1} = 1, \tag{III.B.2.a.ii.4}$$

allowing us to obtain a description of the domain of existence of the phenomenon (III.B.2.a.ii.1). Without loss of generality one can assume that $x_2(0) < x_2(1)$, which implies that $x_3(0) > x_3(1)$; this means that

$$\log\left[\frac{x_2(0)}{x_2(1)}\right] < 0, \qquad \log\left[\frac{x_3(0)}{x_3(1)}\right] > 0.$$

Then conditions (III.B.2.a.ii.4) give

$$\begin{aligned} (\alpha_{22}+1)\log\left[\frac{x_2(0)}{x_2(1)}\right] &+ \alpha_{23}\log\left[\frac{x_3(0)}{x_3(1)}\right] \\ &= \alpha_{32}\log\left[\frac{x_2(0)}{x_2(1)}\right] + (\alpha_{33}+1)\log\left[\frac{x_3(0)}{x_3(1)}\right] = 0. \end{aligned} \tag{III.B.2.a.ii.5}$$

This is impossible if

$$\begin{cases} \alpha_{22} > -1, \\ \alpha_{23} < 0, \end{cases} \text{ or } \begin{cases} \alpha_{22} < -1, \\ \alpha_{23} > 0, \end{cases} \text{ or } \begin{cases} \alpha_{32} > 0, \\ \alpha_{33} < -1, \end{cases} \text{ or } \begin{cases} \alpha_{32} < 0, \\ \alpha_{33} > -1, \end{cases}$$

(see Figure 19). Thus, domains of possible existence of local stable two-period cycles of the type (III.B.2.a.ii.1) are found in the space of parameters, α_{22}, α_{23},

$$\begin{cases} \alpha_{22} < -1, \\ \alpha_{23} < 0, \end{cases} \text{ or } \begin{cases} \alpha_{22} > -1, \\ \alpha_{23} > 0, \end{cases} \tag{III.B.2.a.ii.6}$$

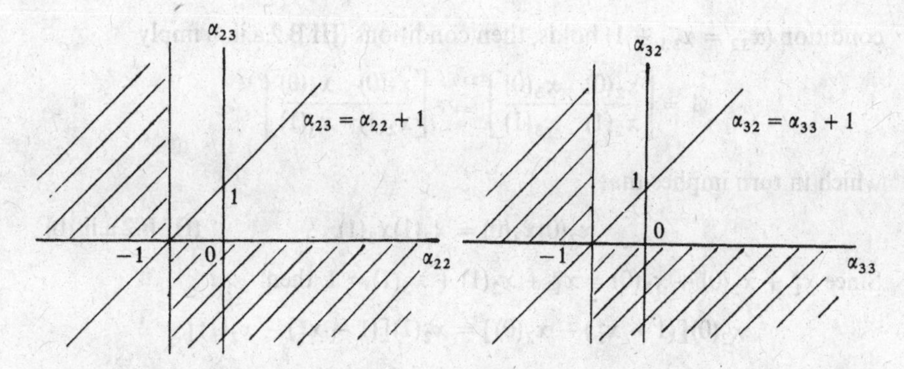

FIGURE 19. Domains (nonshaded areas) of possible existence of local two-period cycles. The straight lines, $\alpha_{23} = \alpha_{22} + 1$, $a_{32} = \alpha_{33} + 1$, present the local periodicity with role reversal phenomenon.

and in the space of parameters, α_{32}, α_{33},

$$\begin{cases}\alpha_{33} < -1, \\ \alpha_{32} < 0,\end{cases} \quad \text{or} \quad \begin{cases}\alpha_{33} > -1, \\ \alpha_{32} > 0.\end{cases} \qquad \text{(III.B.2.a.ii.7)}$$

Another interesting phenomenon in the areas (III.B.2.a.ii.6, 7) is the possible presence of an additional property which demonstrates the existence of a stable two-period cycle with *role reversal*

$$x_2(0) = x_3(1); \qquad x_3(0) = x_2(1). \qquad \text{(III.B.2.a.ii.8)}$$

This phenomenon occurs graphically on the two straight lines

$$\alpha_{23} = \alpha_{22} + 1, \qquad \alpha_{32} = \alpha_{33} + 1. \qquad \text{(III.B.2.a.ii.9)}$$

Next, we will prove that conditions (III.B.2.a.ii.8) and (III.B.2.a.ii.9) are equivalent and will find the explicit analytical formulas for the fixed-point and the two-period cycle (III.B.2.a.ii.1).

If role reversal exists then the conditions (III.B.2.a.ii.4) imply that

$$\left[\frac{x_2(0)}{x_3(0)}\right]^{(\alpha_{22}+1)} \left[\frac{x_3(0)}{x_2(0)}\right]^{\alpha_{23}} = \left[\frac{x_2(0)}{x_3(0)}\right]^{\alpha_{32}} \left[\frac{x_3(0)}{x_2(0)}\right]^{(\alpha_{33}+1)} = 1,$$

or

$$\left[\frac{x_2(0)}{x_3(0)}\right]^{(\alpha_{22}+1-\alpha_{23})} = \left[\frac{x_2(0)}{x_3(0)}\right]^{(\alpha_{32}-\alpha_{33}-1)} = 1.$$

Since $x_2(0) \neq x_2(1) = x_3(0)$, then

$$\alpha_{22} + 1 - \alpha_{23} = \alpha_{32} - \alpha_{33} - 1 = 0.$$

Inversely, if for $\alpha_{23} \neq 0$, the condition $(\alpha_{23} = \alpha_{22} + 1)$ holds; or, if $\alpha_{33} \neq 0$, the condition ($\alpha_{32} = \alpha_{33} + 1$) holds; then conditions (III.B.2.a.ii.4) imply

$$1 = \left[\frac{x_2(0)}{x_2(1)} \frac{x_3(0)}{x_3(1)}\right]^{\alpha_{23}} = \left[\frac{x_2(0)}{x_2(1)} \frac{x_3(0)}{x_3(1)}\right]^{\alpha_{32}},$$

which in turn implies that

$$x_2(0)x_3(0) = x_2(1)x_3(1). \qquad \text{(III.B.2.a.ii.10)}$$

Since $x_1^* + x_2(0) + x_3(0) = x_1^* + x_2(1) + x_3(1) = 1$, then

$$x_2(0)[(1 - x_1^*) - x_2(0)] = x_2(1)[(1 - x_1^*) - x_2(1)],$$

or

$$(1 - x_1^*)[x_2(0) - x_2(1)] = x_2^2(0) - x_2^2(1),$$

or

$$1 - x_1^* = x_2(0) + x_2(1),$$

or

$$1 - x_1^* = x_2(0) + x_3(0) = x_2(1) + x_3(1) = x_2(0) + x_2(1),$$

which immediately implies the "role reversal" phenomenon.

The conditions (III.B.2.a.ii.8) allow for explicitly deriving the analytical form of the fixed-point x_1^* and the components $x_2(0) = x_3(1)$, $x_3(0) = x_2(1)$ of the stable two-period cycle. The role reversal conditions invert the expressions (III.B.2.a.ii.2) into

$$\frac{x_i(0)}{x_1^*} = A_i[x_1^*]^{\alpha_{i1}}[x_3(0)]^{\alpha_{i2}}[x_2(0)]^{\alpha_{i3}}, \qquad i = 2, 3, \qquad \text{(III.B.2.a.ii.11)}$$

or

$$\begin{cases} 1 = A_2[x_1^*]^{\alpha_{21}+1}[x_2(0)]^{\alpha_{23}-1}[x_3(0)]^{\alpha_{22}}, \\ 1 = A_3[x_1^*]^{\alpha_{31}+1}[x_2(0)]^{\alpha_{33}}[x_3(0)]^{\alpha_{32}-1}, \end{cases}$$

or, if

$$y = x_2(0)x_3(0), \qquad \text{(III.B.2.a.ii.12)}$$

then

$$\begin{cases} 1 = A_2[x_1^*]^{\alpha_{21}+1}y^{\alpha_{22}}, \\ 1 = A_3[x_1^*]^{\alpha_{31}+1}y^{\alpha_{33}}. \end{cases} \qquad \text{(III.B.2.a.ii.13)}$$

By taking the logarithms one obtains a system of two linear equations for $\log x_1^*$, $\log y$

$$\begin{cases} (\alpha_{21} + 1) \log x_1^* + \alpha_{22} \log y = -\log A_2, \\ (\alpha_{31} + 1) \log x_1^* + \alpha_{33} \log y = -\log A_3, \end{cases} \qquad \text{(III.B.2.a.ii.14)}$$

with a determinant Δ_1 given by

$$\Delta_1 = \begin{vmatrix} \alpha_{21} + 1 & \alpha_{22} \\ \alpha_{31} + 1 & \alpha_{33} \end{vmatrix}. \qquad \text{(III.B.2.a.ii.15)}$$

If Δ_1 is nonzero, then the solution of (III.B.2.a.ii.14) is

$$x_1^* = \left[\frac{A_3^{\alpha_{22}}}{A_2^{\alpha_{33}}}\right]^{1/\Delta_1}, \qquad y = \left[\frac{A_2^{\alpha_{31}+1}}{A_3^{\alpha_{21}+1}}\right]^{1/\Delta_1}. \qquad \text{(III.B.2.a.ii.16)}$$

Further, for evaluating $x_2(0)$, $x_3(0)$, we have two conditions

$$x_2(0) + x_3(0) = 1 - x_1^*,$$

$$x_2(0) + x_3(0) = y,$$

which produce the following quadratic equation for $x_2(0)$, $x_3(0)$,

$$Z^2 - (1 - x_1^*)Z + y = 0, \qquad \text{(III.B.2.a.ii.17)}$$

with a solution

$$Z_{1,2} = \frac{1 - x_1^*}{2} \pm \sqrt{\left(\frac{1 - x_1^*}{2}\right)^2 - y}. \qquad \text{(III.B.2.a.ii.18)}$$

If

$$\Delta_1 = 0 \quad \text{and} \quad A_2^{\alpha_{33}} \neq A_3^{\alpha_{22}}, \qquad \text{(III.B.2.a.ii.19)}$$

the system (III.B.2.a.ii.14) has no solutions. If, on the other hand,

$$\Delta_1 = 0 \quad \text{and} \quad A_2^{\alpha_{33}} = A_3^{\alpha_{22}}, \qquad \text{(III.B.2.a.ii.20)}$$

then the system (III.B.2.a.ii.13) degenerates to one equation

$$1 = A_2[x_1^*]^{\alpha_{21}+1} y^{\alpha_{22}}.$$

Choosing an arbitrary x_1^* in the domain (0, 1) and from

$$y = A_2^{-1/\alpha_{22}}[x_1^*]^{-(\alpha_{21}+1)/\alpha_{22}},$$

we obtain an infinite number of solutions in association with the solution of the quadratic equation (III.B.2.a.ii.17). Figure 20 shows the qualitative plot of the relative one-stock, three-location dynamics with the stable partial attractor x_1^* and stable local two-periodic role reversal cycle. At the lower right side of this figure the exponent matrix (α_{ij}) responsible for the phenomenon is shown.

The existence of a stable partial attractor x_1^* with the property that

$$\lim_{t\to\infty} x_1(t + 1) = x_1^*$$

allows us to transfer all considerations of the one-stock, two-location relative dynamics into the framework of the one-stock, three-location model. This is so because the approximative conservation condition

$$x_2(t) + x_3(t) \simeq 1 - x_1^*$$

holds at the vicinity of x_1^*.

We obtain in the one-stock, three-location model, all the phenomena found in the two-location case including the three fundamental bifurcations of

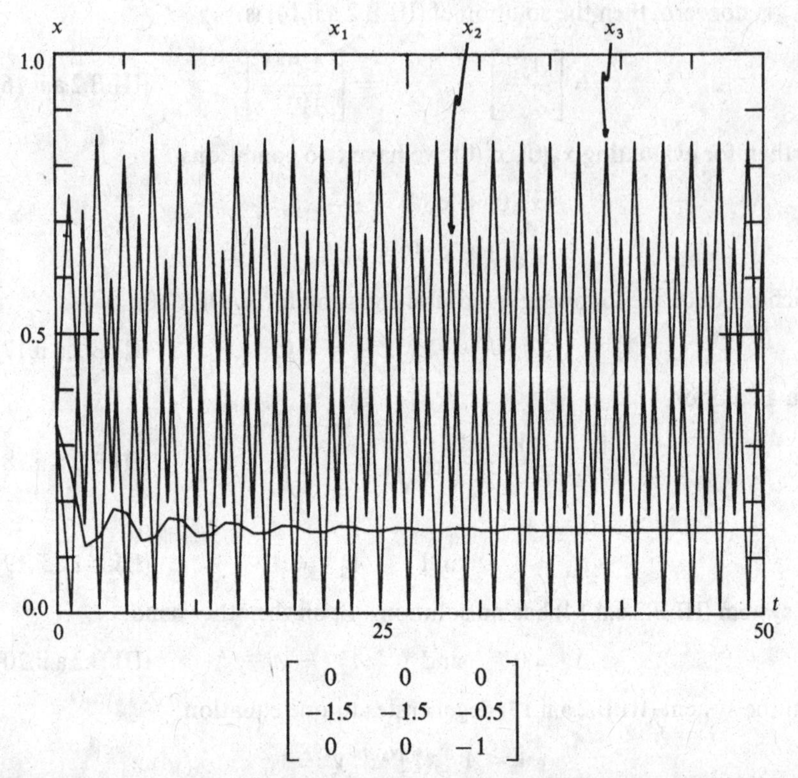

FIGURE 20. A fixed point and a stable two-period cycle in the three-location, one-stock case: $A_1 = 1$, $A_2 = 10^{-4}$, $A_3 = 10$. Reprinted by permission of the publisher from "The Onset of Turbulence in Discrete Relative Multiple Spatial Dynamics" by D.S. Dendrinos and M. Sonis, *Journal of Applied Mathematics and Computation*, 22:25–44, Copyright 1987 by Elsevier Science Publishing Co., Inc.

discrete dynamics, which are the Hopf-like discrete bifurcation, period-doubling behavior, and the pathway to deterministic chaos through the Feigenbaum slope-sequences. We leave the proper analytical treatment of these propositions to the interested reader. In Figure 21 the case of a stable three-period cycle for all three locations is shown, where two locations (stocks x_1 and x_3) experience role-reversal, shifting from almost extinction to almost complete dominance.

Absence of stable partial attractors generates much more complicated and interesting phenomena, including "strange attractors" and "containers," to be demonstrated graphically next.

iii. Strange Attractors and Containers

An example of a "strange container," meaning an area or a curve in the state variables' space ($0 \leq x_1, x_2, x_3 \leq 1$) entered and spanned by the iterative process without leaving it, is shown in Figure 22. Once trapped inside its

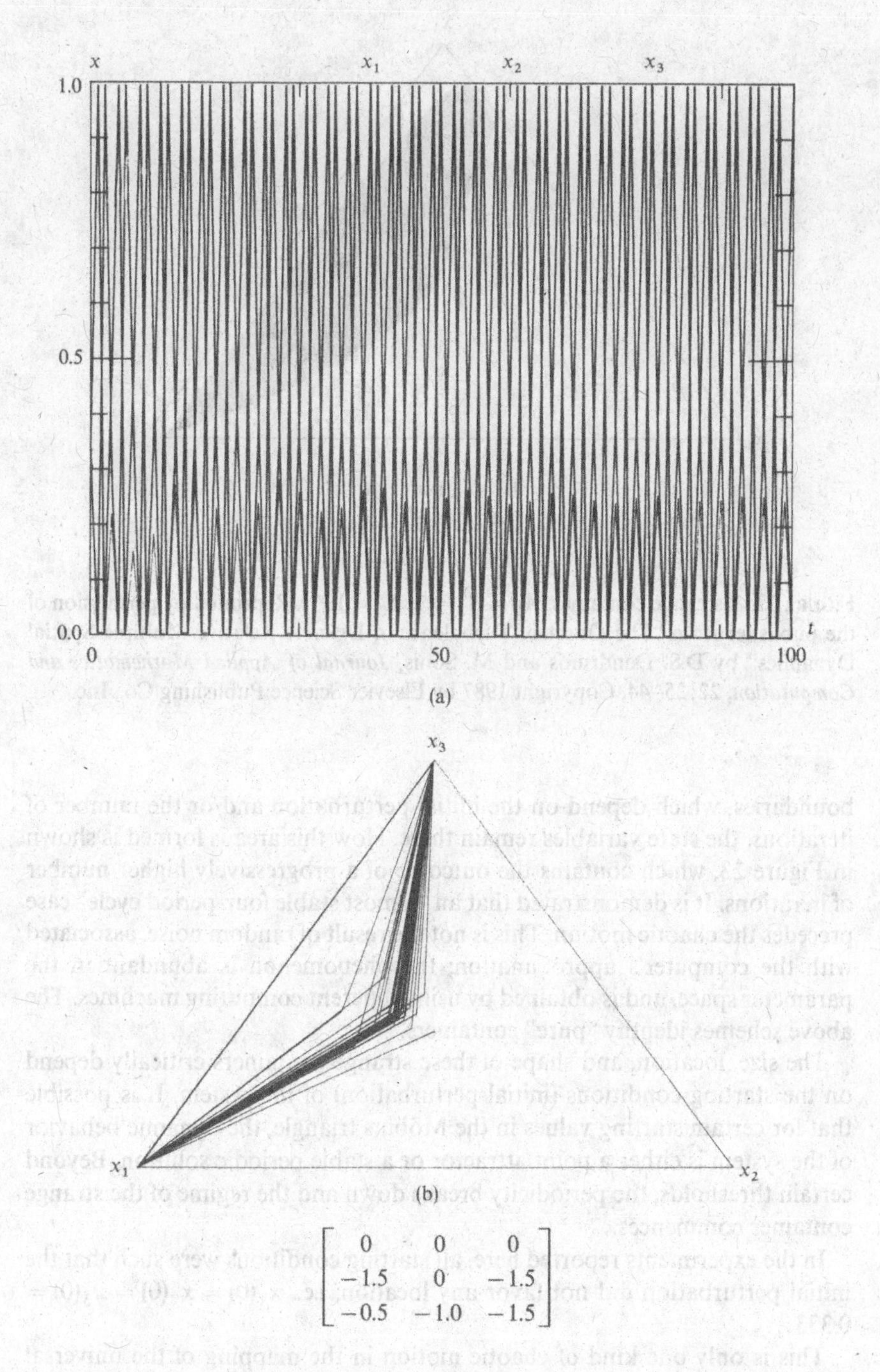

$$\begin{bmatrix} 0 & 0 & 0 \\ -1.5 & 0 & -1.5 \\ -0.5 & -1.0 & -1.5 \end{bmatrix}$$

FIGURE 21. A stable three-period cycle: $A_1 = 1$, $A_2 = A_3 = 10^{-4}$.

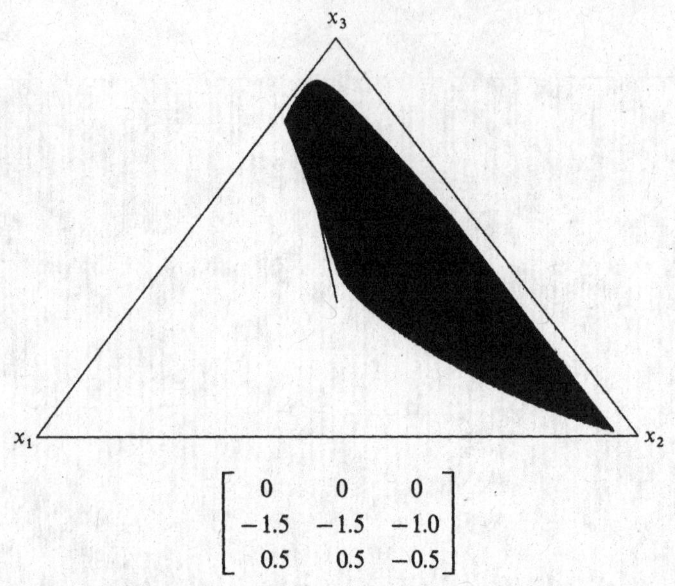

FIGURE 22. A strange container: $A_1 = A_3 = 1$, $A_2 = 10^{-3}$. Reprinted by permission of the publisher from "The Onset of Turbulence in Discrete Relative Multiple Spatial Dynamics" by D.S. Dendrinos and M. Sonis, *Journal of Applied Mathematics and Computation*, 22:25–44, Copyright 1987 by Elsevier Science Publishing Co., Inc.

boundaries, which depend on the initial perturbation and/or the number of iterations, the state variables remain there. How this area is formed is shown in Figure 23, which contains the outcome of a progressively higher number of iterations. It is demonstrated that an "almost stable four-period cycle" case precedes the chaotic motion. This is not the result of random noise, associated with the computer's approximation; the phenomenon is abundant in the parameter space, and is obtained by using different computing machines. The above schemes identify "pure" containers.

The size, location, and shape of these strange containers critically depend on the starting conditions (initial perturbation) of the system. It is possible that for certain starting values in the Möbius triangle, the dynamic behavior of the system is either a point attractor or a stable periodic solution. Beyond certain thresholds, the periodicity breaks down and the regime of the strange container commences.

In the experiments reported here, all starting conditions were such that the initial perturbation did not favor any location; i.e., $x_1(0) = x_2(0) = x_3(0) = 0.333\ldots$.

This is only one kind of chaotic motion in the mapping of the universal discrete relative dynamics. More cases of this phenomenon are provided in Dendrinos and Sonis (1987).

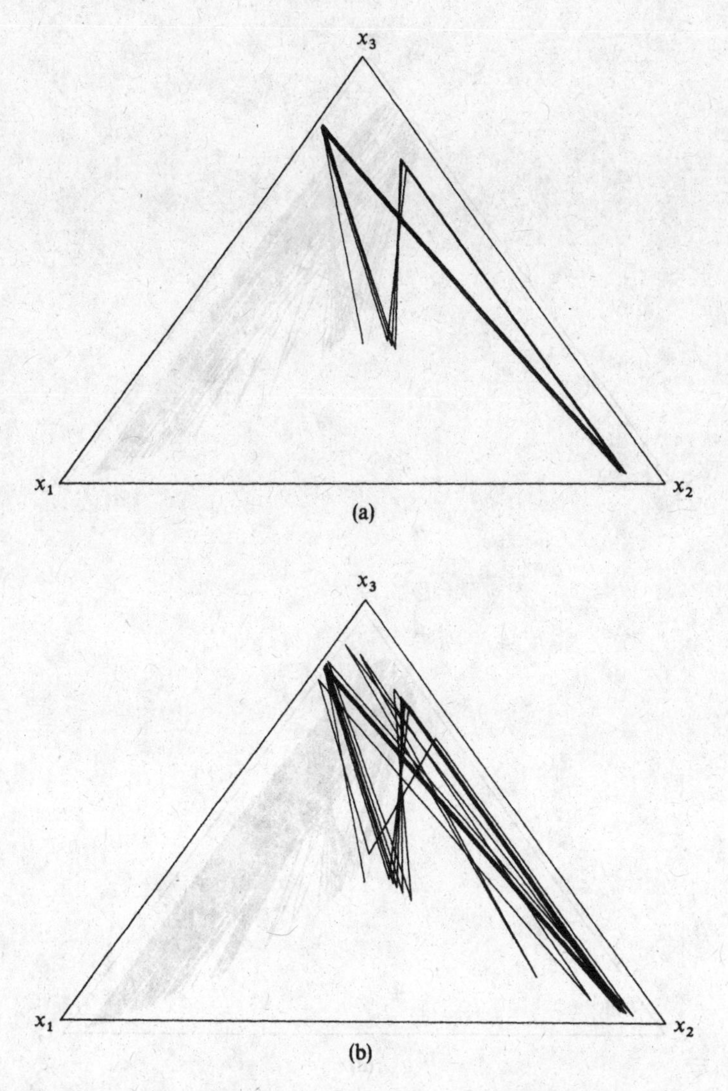

FIGURE 23. The formation of a strange container; specifications identical to those in Figure 22.

iv. Hybrid Containers, Attractors

Two "hybrid container–attractor" cases are shown in Figures 24 and 25. They demonstrate the existence of hybrid configurations with different sizes. In Figure 24 the area in which the state variables' range is contained spans a large part of the feasible space; whereas in Figure 25 this space is confined to a small area and close to the repulsion region drawing a thin ring around it.

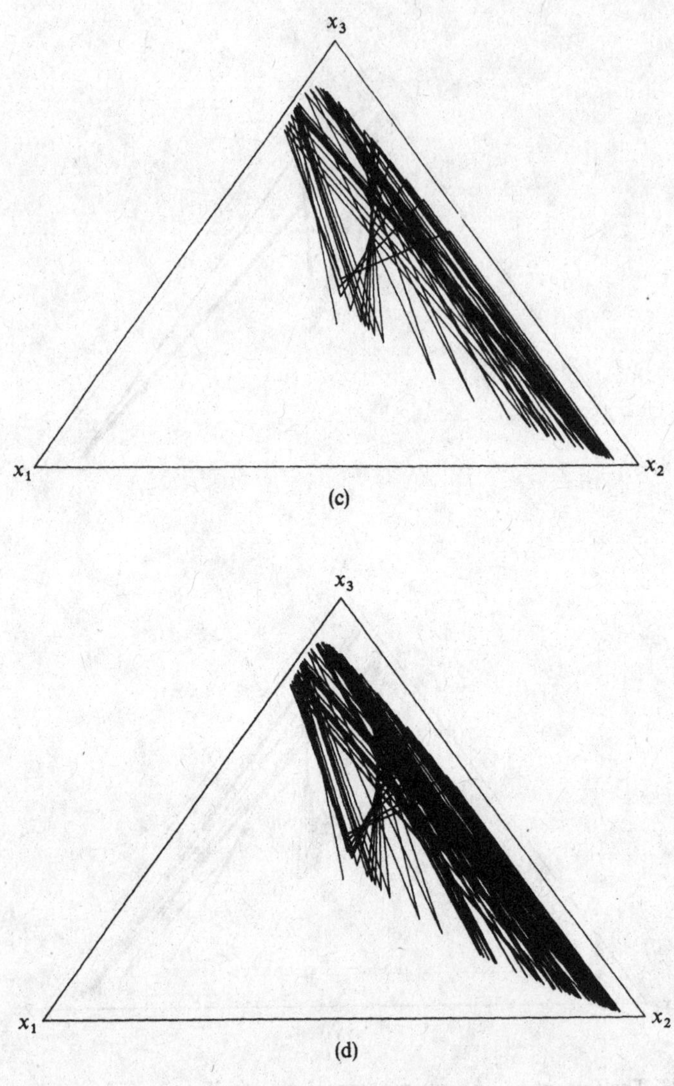

FIGURE 23 (*continued*)

v. The Onset of Turbulence in the Three-Location, One-Stock Model

Whereas the two-location, one-stock model closely resembles the May process in its turbulent movement from a fixed point to chaotic motion (although not following Feigenbaum's universal numbers on the way to turbulence), the three-location, one-stock case has surprises for us. There are no continuous transitions depicting cascades of 2^k $(k \to \infty)$ stable period cycles, as critical values of the bifurcation parameters A are crossed, with windows of $n2^k$ periodic motion.

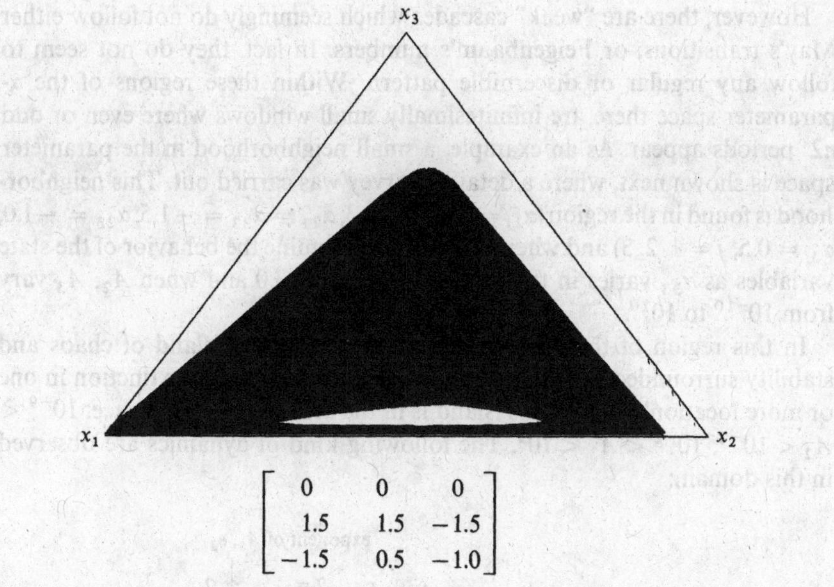

FIGURE 24. A relatively large container–attractor: $A_1 = 1$, $A_2 = 0.1$, $A_3 = 10^{-3}$. Reprinted by permission of the publisher from "The Onset of Turbulence in Discrete Relative Multiple Spatial Dynamics" by D.S. Dendrinos and M. Sonis, *Journal of Applied Mathematics and Computation*, 22:25–44, Copyright 1987 by Elsevier Science Publishing Co., Inc.

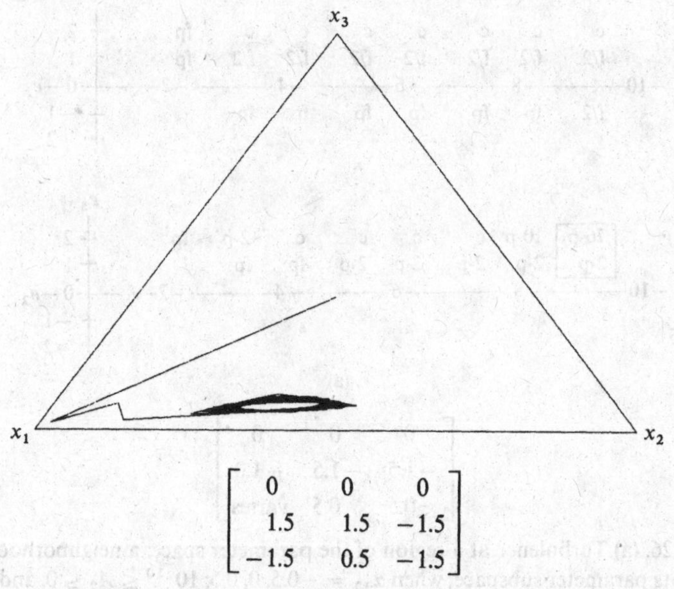

FIGURE 25. An attraction ring: $A_1 = 1$, $A_2 = 0.1$, $A_3 = 10^{-3}$.

However, there are "weak" cascades which seemingly do not follow either May's transitions, or Feigenbaum's numbers. In fact, they do not seem to follow any regular or discernible pattern. Within these regions of the α-parameter space there are infinitesimally small windows where even or odd $n2^k$ periods appear. As an example, a small neighborhood in the parameter space is shown next, where a detailed survey was carried out. This neighborhood is found in the region ($\alpha_{1j} = 0, j = 1, 2, 3; \alpha_{21} = \alpha_{22} = -1.5; \alpha_{23} = -1.0;$ $\alpha_{3j} = 0.5, j = 1, 2, 3$) and where $A_1 = 1$. We examine the behavior of the state variables as α_{33} varies in the area $-0.5 \leq \alpha_{33} \leq 3.0$ and when A_2, A_3 vary from 10^{-10} to 10^{10}.

In this region of the parameter space one finds an island of chaos and stability surrounded by unstable states, i.e., convergence to extinction in one or more location's stock. This island is in the area of the A subspace: $10^{-9} < A_2 < 10^{-2}$, $10^{-2} < A_3 < 10^2$. The following kind of dynamics are observed in this domain:

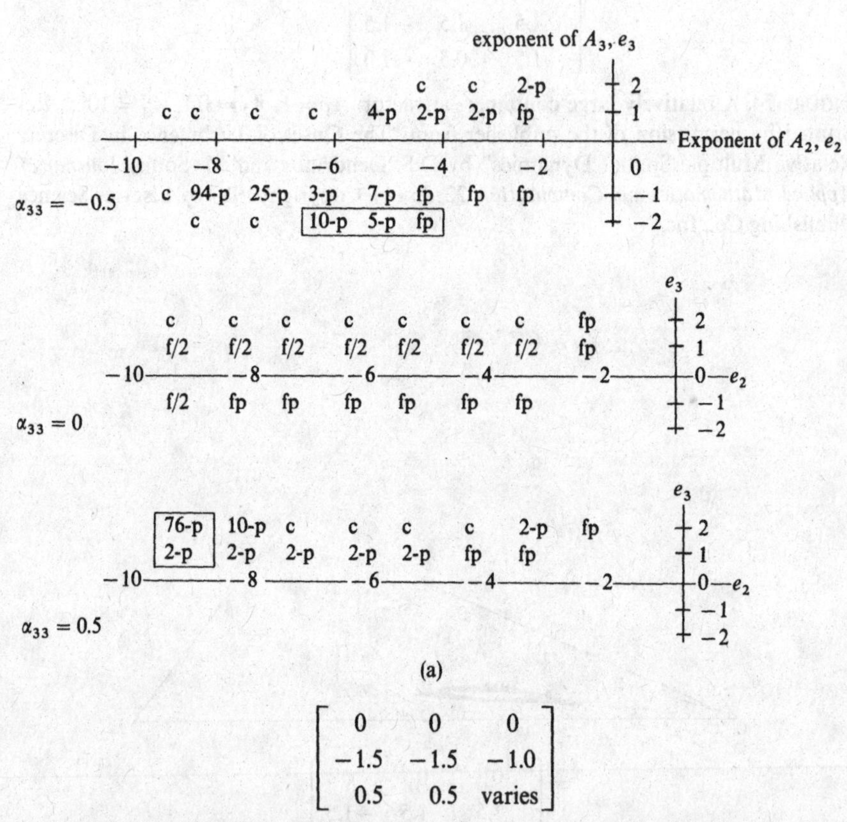

$$\begin{bmatrix} 0 & 0 & 0 \\ -1.5 & -1.5 & -1.0 \\ 0.5 & 0.5 & \text{varies} \end{bmatrix}$$

FIGURE 26. (a) Turbulence at a region of the parameter space: a neighborhood in the exponents parameter subspace, when $\alpha_{33} = -0.5, 0, 0.5$; $10^{-10} \leq A_2 \leq 0$; and $10^{-2} \leq A_3 \leq 10^2$. Designations: f/2 stands for one fixed point and two stable two-period cycles; fp stands for fixed point; k-p is a stable k-period cycle; and c stands for chaotic motion; nondesignated intersections imply unstable behavior.

unstable states, where one state variable converges to zero—designated as 10^{-18} in the computing algorithm—and the other two converge to non-zero and nonidentical or identical values;
fixed points;
partial turbulence, where one location's stock converges to a fixed point, whereas the other two converge to a stable two-period cycle with role reversal;
various stable periodic motions of the type 2^k, and/or $n2^k$;
chaos.

In Figure 26(a), (b) the various behaviors are shown corresponding to a step $\lambda = 10$ in the A's, and a step $\mu = 0.5$ in α_{33}. Undesignated points on the grid

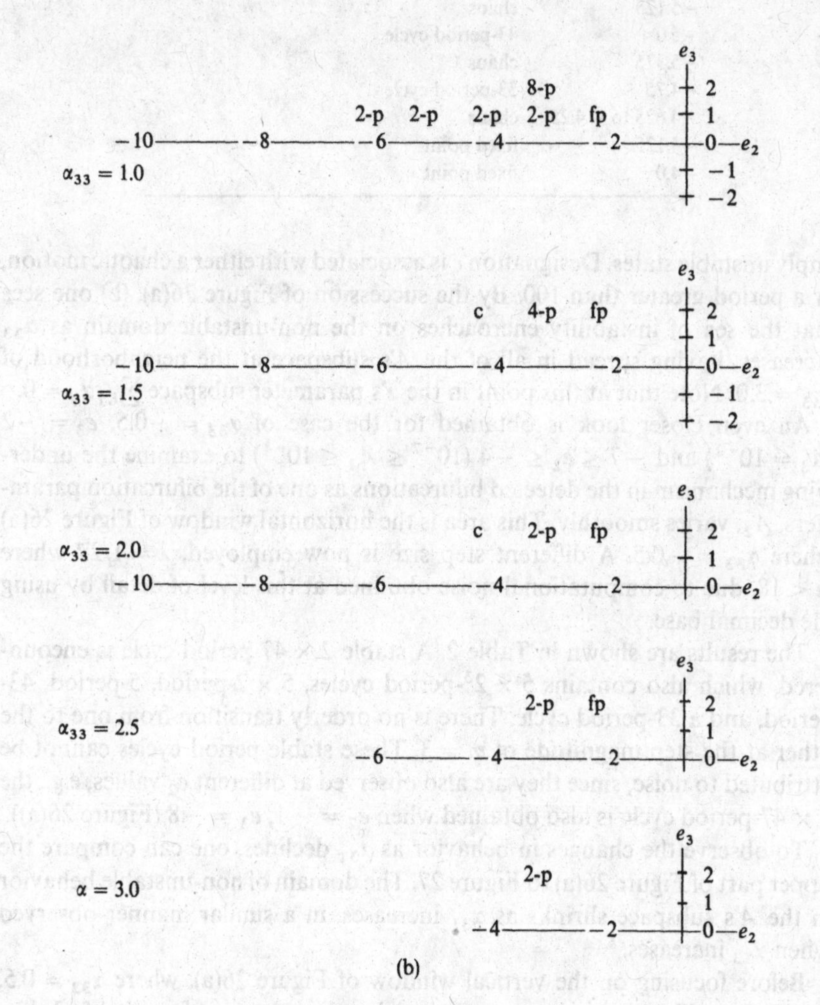

FIGURE 26. (b) Turbulence at a region of the parameter space: specifications are the same as in Figure 26(a), and $\alpha_{33} = 1, 1.5, 2, 2.5, 3$.

TABLE 2. A close look at the turbulent regime of a particular region of the exponents' (structural) parameter subspace: $e_3 = -2$; $-4 \leq e_2 \leq -7$, and

$$[\alpha_{ij}] = \begin{bmatrix} 0 & 0 & 0 \\ -1.5 & -1.5 & -1.0 \\ 0.5 & 0.5 & -0.5 \end{bmatrix}.$$

e_2	Dynamic behavior type
−7.0	94-period (2 × 47-period) cycle
−6.875 to −6.50	chaos (at −6.50 a possible 50-period cycle)
−6.375	40-period (5 × 2^3-period) cycle
−6.25	10-period (5 × 2-period) cycle
−6.125 to −5.25	5-period cycle
−5.125	chaos
−5.0	43-period cycle
−5.375	chaos
−4.75	33-period cycle
−4.625 to −4.25	chaos
−4.125	fixed point
−4.0	fixed point

imply unstable states. Designation c is associated with either a chaotic motion, or a period greater than 100. By the succession of Figure 26(a), (b) one sees that the sea of instability encroaches on the non-unstable domain as α_{33} increases, having spread in all of the A's subspace at the neighborhood of $\alpha_{33} = 3.0$. Note that at this point in the α's parameter subspace $\sum_{i,j}\alpha_{ij} = 0$.

An even closer look is obtained for the case of $\alpha_{33} = -0.5$, $e_3 = -2$ ($A_3 = 10^{-2}$) and $-7 \leq e_2 \leq -4$ ($10^{-7} \leq A_2 \leq 10^{-4}$) to examine the underlying mechanism in the detected bifurcations as one of the bifurcation parameters, A_2, varies smoothly. This area is the horizontal window of Figure 26(a) where $\alpha_{33} = -0.5$. A different step size is now employed, $\lambda = 1/2^m$ where $m < 18$, due to computational noise obtained at this level of detail by using the decimal base.

The results are shown in Table 2. A stable 2 × 47-period cycle is encountered, which also contains 5 × 2^3-period cycles, 5 × 2-period, 5-period, 43-period, and a 33-period cycle. There is no orderly transition from one to the other at the step magnitude of $m = 3$. These stable period cycles cannot be attributed to noise, since they are also observed at different e_2 values, e.g., the 2 × 47-period cycle is also obtained when $e_2 = -1$, $e_3 = -8$ (Figure 26(a)).

To observe the changes in behavior as α_{31} declines, one can compare the upper part of Figure 26(a) to Figure 27. The domain of non-unstable behavior in the A's subspace shrinks as α_{31} increases, in a similar manner observed when α_{33} increases.

Before focusing on the vertical window of Figure 26(a), where $\alpha_{33} = 0.5$, which, as will be seen later, contains the closest case of a pitchfork-like bifurcation cascade similar to May's, another window will be presented, shown

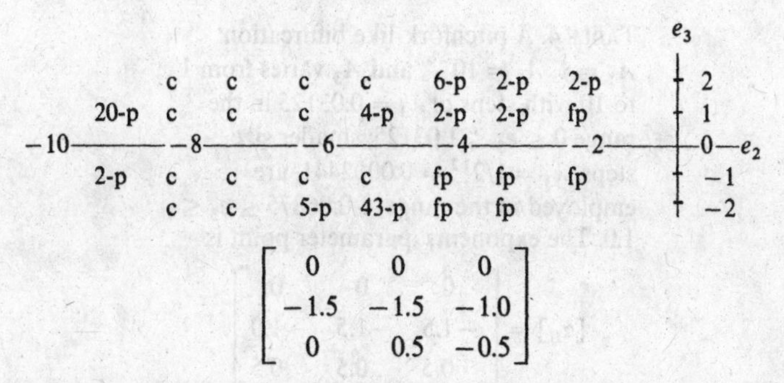

$$\begin{bmatrix} 0 & 0 & 0 \\ -1.5 & -1.5 & -1.0 \\ 0 & 0.5 & -0.5 \end{bmatrix}$$

FIGURE 27. Further evidence of turbulence in the one-stock, three-location model: the behavior is shown when $-10 \leq e_2 \leq -2$ and $-2 \leq e_3 \leq 2$. Compare this with the upper part of Figure 26(a).

in Table 3. This window shows how a 3×2^2-period cycle is transformed into a 7×2-period cycle through seemingly different sequences of chaotic motion. The step size λ_3 is equal to $1/2^m$ where $m = 8$ in this case.

How a pitchfork-like bifurcation scheme appears in some areas of the parameter space in the three-location, one-stock model is demonstrated in Table 4. From $e_3 = 0\,(A_3 = 1)$ and up to $e_3 = 0.65$ the system exhibits a stable two-period cycle; around this threshold the stable two-period cycle is transformed into a stable four-period cycle lasting to approximately 0.9375, where it is expectedly transformed into a stable eight-period cycle. The next series of transformations are picked up by the lower step size, as indicated in Table 4, which employs a finer step size of $\lambda_3 = 1/2^{12} = 0.0002441$.

We only looked for up to 128-period cycles (having examined the last 200 steps of 5000 iterations) with the first step size. No windows of $n2^k$ stable period cycles were detected; and no evidence was found that as A_3 approaches the chaotic regime odd-period cycles start to appear.

TABLE 3. Some other type of chaotic motion: the transformation of a stable 12-period cycle into a stable 14-period cycle; $A_1 = 1$, $A_2 = 10^{-6}$; parameter A_3 varies between 10^{-2} and 1, with steps of $10^{-0.125}$ (or $1/2^8$). The exponents' space point is at

$$[\alpha_{ij}] = \begin{bmatrix} 0 & 0 & 0 \\ -1.5 & 0 & 1.5 \\ 0 & 1.0 & -1.5 \end{bmatrix}.$$

e_3	Dynamic behavior type
−2.0	12-period (3×2^2-period) cycle
−1.875 to −0.125	chaos
0	14-period (2×7-period) cycle

TABLE 4. A pitchfork-like bifurcation: $A_1 = 1$, $A_2 = 10^{-9}$, and A_3 varies from 1 to 10 with steps of $\lambda_3 = 0.03125$ in the range $0 \leq e_3 \leq 1.03125$; smaller size steps, $\lambda_3 = 1/2^{12} = 0.0002441$, are employed in the range $-0.96875 \leq e_3 \leq 1.0$. The exponents' parameter point is

$$[\alpha_{ij}] = \begin{bmatrix} 0 & 0 & 0 \\ -1.5 & -1.5 & -1.0 \\ 0.5 & 0.5 & 0.5 \end{bmatrix}.$$

e_3	Dynamic behavior type
0 to 0.625	2-period cycle
0.65625 to 0.90625	4-period cycle
0.9375	8-period cycle
0.96875	16-period cycle
0.9785	32-period cycle
0.98096	64-period cycle
0.9833984	128-period cycle
1.03125	chaos

A note is in order regarding these searches into the parameter space: to distinguish between 2^k and 2^{k+1} cycles the following criterion was used: if at 18 decimals the $x(t + 2^{k+1})$ and $x(t)$ iterations had more identical decimals than the $x(t + 2^k)$ and $x(t)$ iterates, then a 2^{k+1}-period cycle is assumed to have occurred.

A major finding from the universal discrete relative dynamics model is that the May chaos is only one type among many other processes giving rise to an aperiodic (but deterministic) motion following a succession of periodic movements. Further, there is some (deterministic) mechanism underlying inter- (as well as intra-) chaotic transformations. Another major finding from this universal discrete relative dynamics model is that Li and Yorke's result "period three implies chaos" (1975), to the extent that it requires the three-period cycle to immediately precede chaos, does not generally hold. This topic is the subject of forthcoming work.

vi. Higher Iterates

In discrete maps of socio-spatial dynamics, higher iterates identify an important event in the spatiotemporal choice process depicted by the universal map. They depict time-delay effects underlying spatiotemporal choices. Specifically, they identify intertemporal, interspatial, and interstock response functions as resulting from a variety of factors which may affect these choices. Some of these factors may be related to various spatiotemporal transaction costs, including transportation costs.

A key finding from the analysis which follows is that order is found in these higher iterates, even though, in certain cases, the underlying time-one map is chaotic.

In Figure 28 the $t+1$, $t+2$, $t+3$, $t+4$, $t+5$, and $t+6$ iterates of $x_1(t+n)$ versus $[x_1(t), x_3(t)]$ are shown. The specific case is that of a fixed point. In Figure 29 the first and second iterate of x_1 and x_2 are plotted against x_2 and x_3 for the case of a fixed point. Figure 30 depicts the same iterates for a two-period cycle. Finally, Figure 31 identifies a chaotic case, whereas Figure 32 depicts a stable four-period cycle. Note that the cases in Figures 29, 30, 31, and 32 are in the same neighborhood of α_{33} and A_2; thus, an indication is provided of the changes in the surfaces of higher iterates as α_{33} or A_2 change.

From Figure 31, and in the case of $x_2(t+1)$ versus $[x_2(t), x_3(t)]$, one notes that the behavior of x_2 depends (as discussed in previous sections) on the intersection of the three-dimensional plane $x_2(t+1) = x_2(t)$ with the surface of the $x_2(t+1)$ iterate for any value $x_3(t)$. Each point of intersection has a slope in the $[x_2(t+1), x_2(t)]$ plane, and the behavior of x_2 depends on that slope: if it is less than one in absolute value x_2 exhibits stable behavior, otherwise it is either periodic or chaotic.

b. The Four-Location, One-Stock Case

i. Some General Observations

In the four-dimensional case, local and partial turbulence demonstrates itself in two ways: either by one location exhibiting fixed-point behavior, while the other three are in a stable two-period cycle; or by two locations being at fixed-point motion, whereas the other two are demonstrating a stable two-period cycle with role reversal (due to the conservation condition). In Figure 33(a) stable periodic motion cases in all four locations are shown. Examples of fixed-point and stable two-period cycles combinations are given in Figure 33(b) and 33(c).

Parameter values responsible for the results shown in Figures 33(a), (b), (c) are as follows:

for Figure 33(a): $A_1 = A_3 = 1$, $A_2 = 10^{-4}$, $A_4 = 10^2$, and

$$[\alpha_{ij}] = \begin{bmatrix} 0 & 0 & 0 & 0 \\ -1.5 & -1.5 & -1.5 & -1.5 \\ -1.5 & 0.5 & -0.5 & 1.5 \\ 0.5 & 0.5 & 0.5 & 0.5 \end{bmatrix};$$

for Figure 33(b) (one fixed point, three stable two-period cycles): $A_1 = 1$, $A_2 = A_3 = A_4 = 10^{-4}$, and

$$[\alpha_{ij}] = \begin{bmatrix} 0 & 0 & 0 & 0 \\ -1.5 & -1.5 & -1.5 & -1.5 \\ -1.5 & -0.5 & -1.5 & -1.5 \\ -0.5 & -1.5 & -1.5 & -1.5 \end{bmatrix};$$

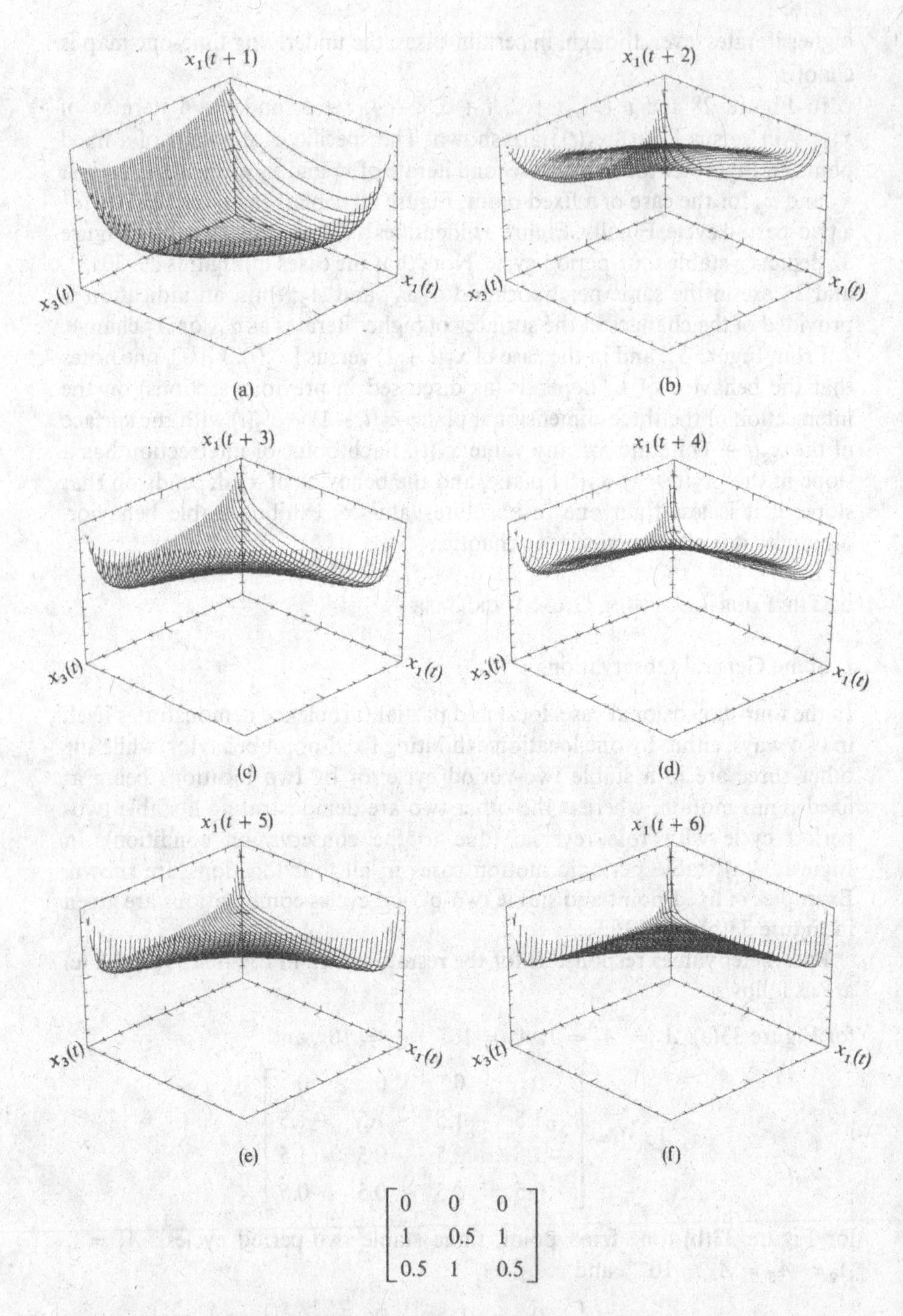

$$\begin{bmatrix} 0 & 0 & 0 \\ 1 & 0.5 & 1 \\ 0.5 & 1 & 0.5 \end{bmatrix}$$

FIGURE 28. The higher iterates of x_1 plotted against x_1 and x_3 in the case of a fixed point: $A_1 = A_3 = 1$, $A_2 = 10^2$.

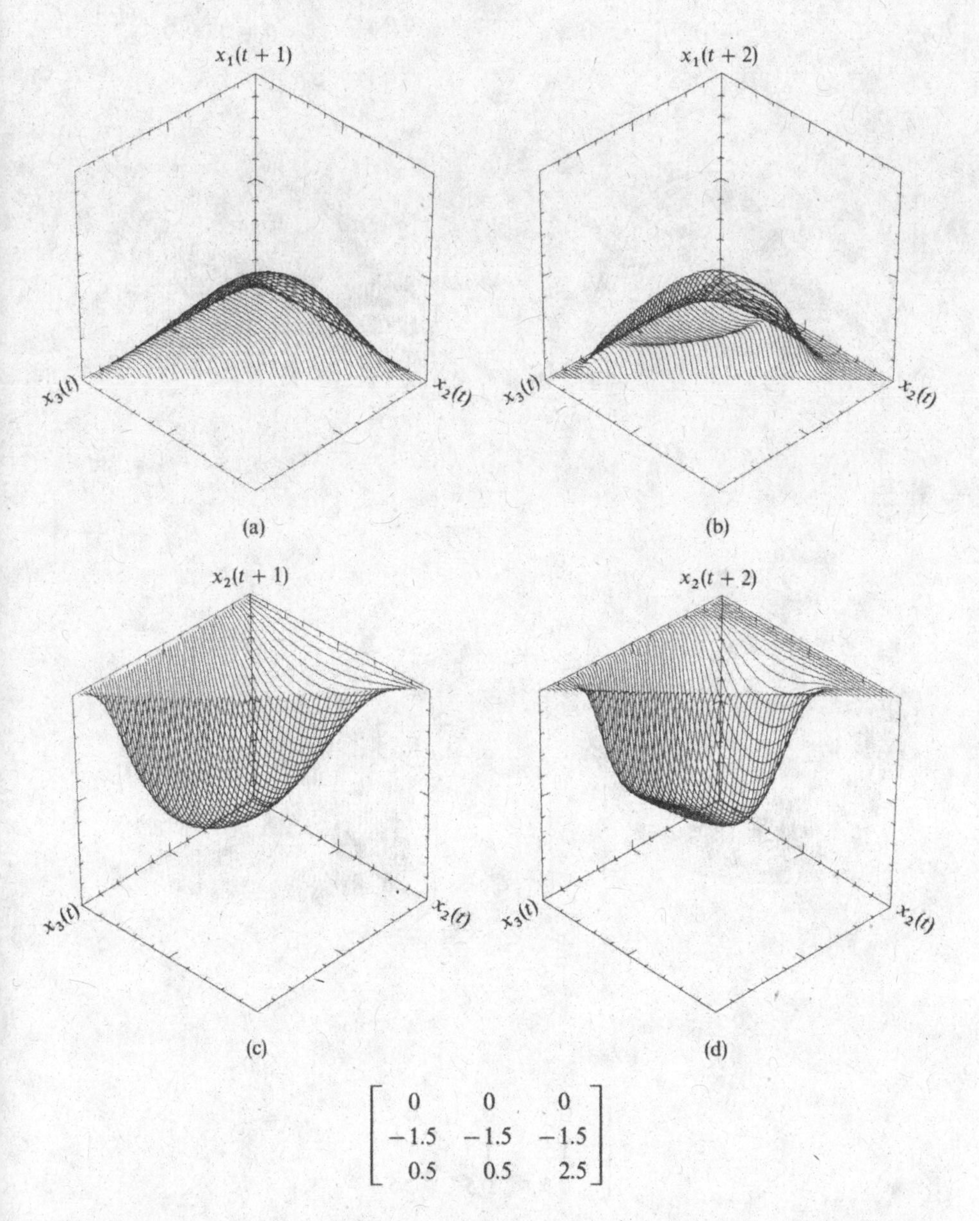

$$\begin{bmatrix} 0 & 0 & 0 \\ -1.5 & -1.5 & -1.5 \\ 0.5 & 0.5 & 2.5 \end{bmatrix}$$

FIGURE 29. The first and second iterates of x_1 and x_2 plotted against x_2 and x_3 for a fixed point: $A_1 = 1$, $A_2 = 10^{-2}$, $A_3 = 10^2$.

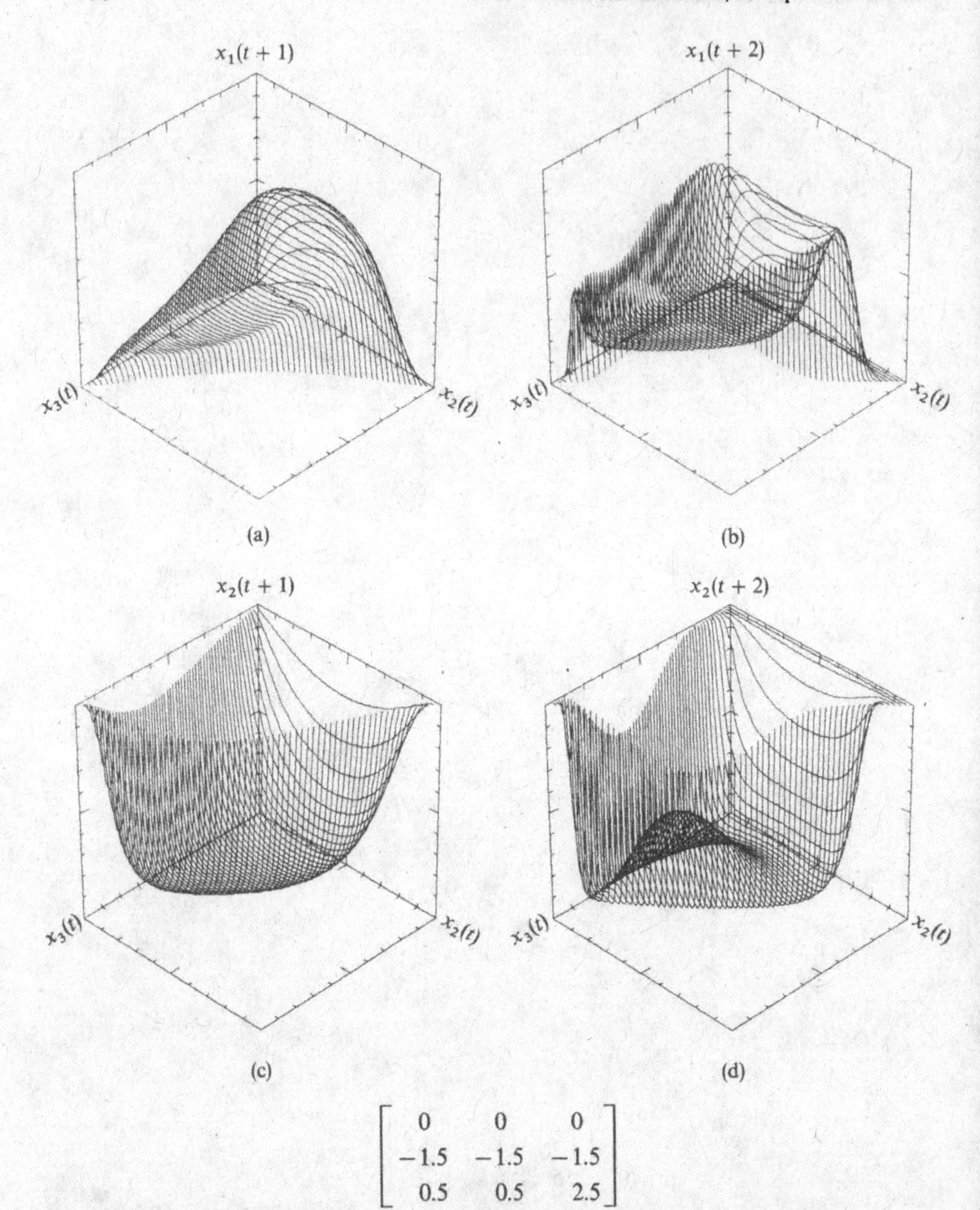

FIGURE 30. The first and second iterates of x_1 and x_2 plotted against x_2 and x_3 for a stable two-period cycle: $A_1 = 1$, $A_2 = 10^{-3}$, $A_3 = 10^2$.

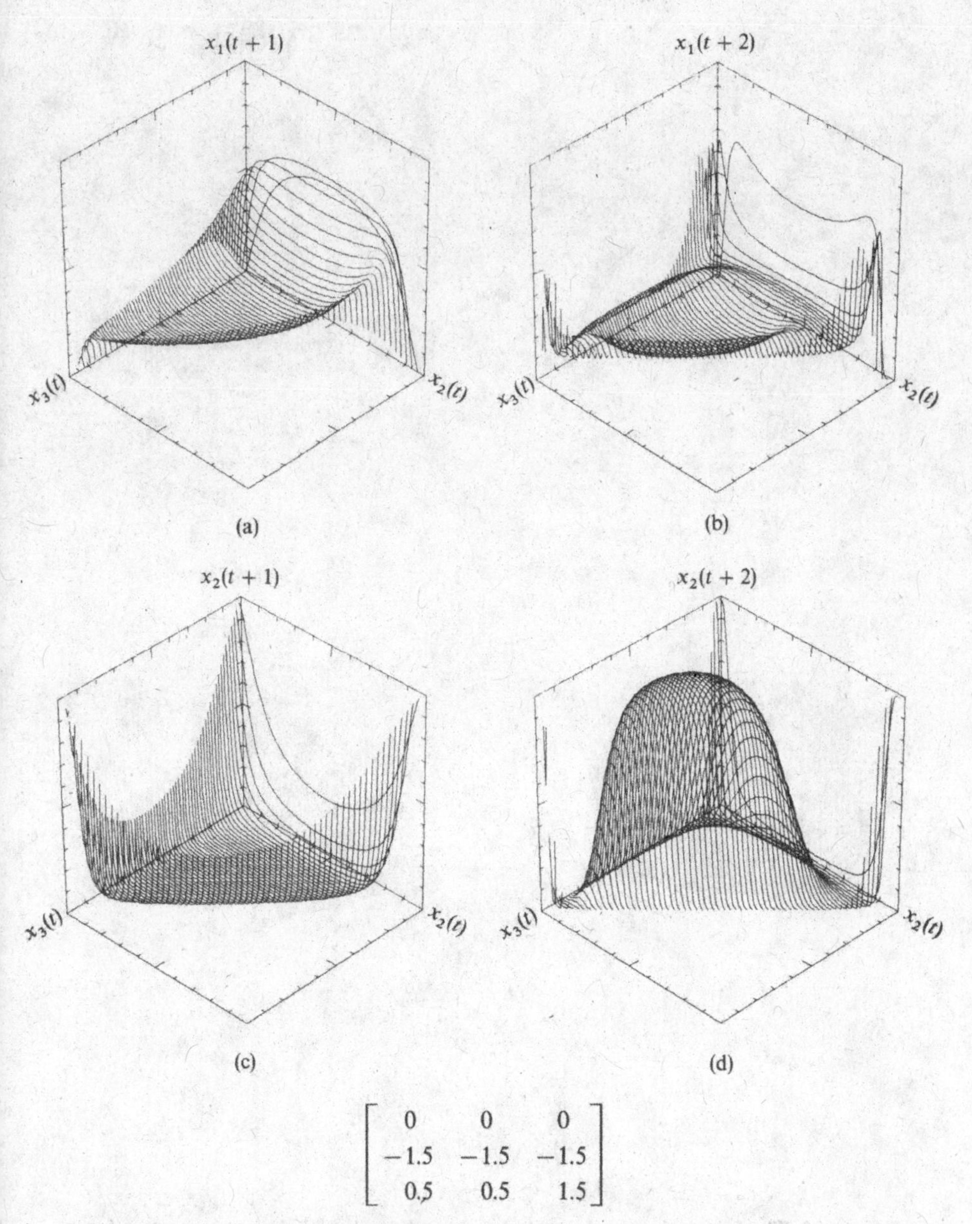

$$\begin{bmatrix} 0 & 0 & 0 \\ -1.5 & -1.5 & -1.5 \\ 0.5 & 0.5 & 1.5 \end{bmatrix}$$

FIGURE 31. The first and second iterates of x_1 and x_2 plotted against x_2 and x_3 for chaos: $A_1 = 1$, $A_2 = 10^{-4}$, $A_3 = 10^2$.

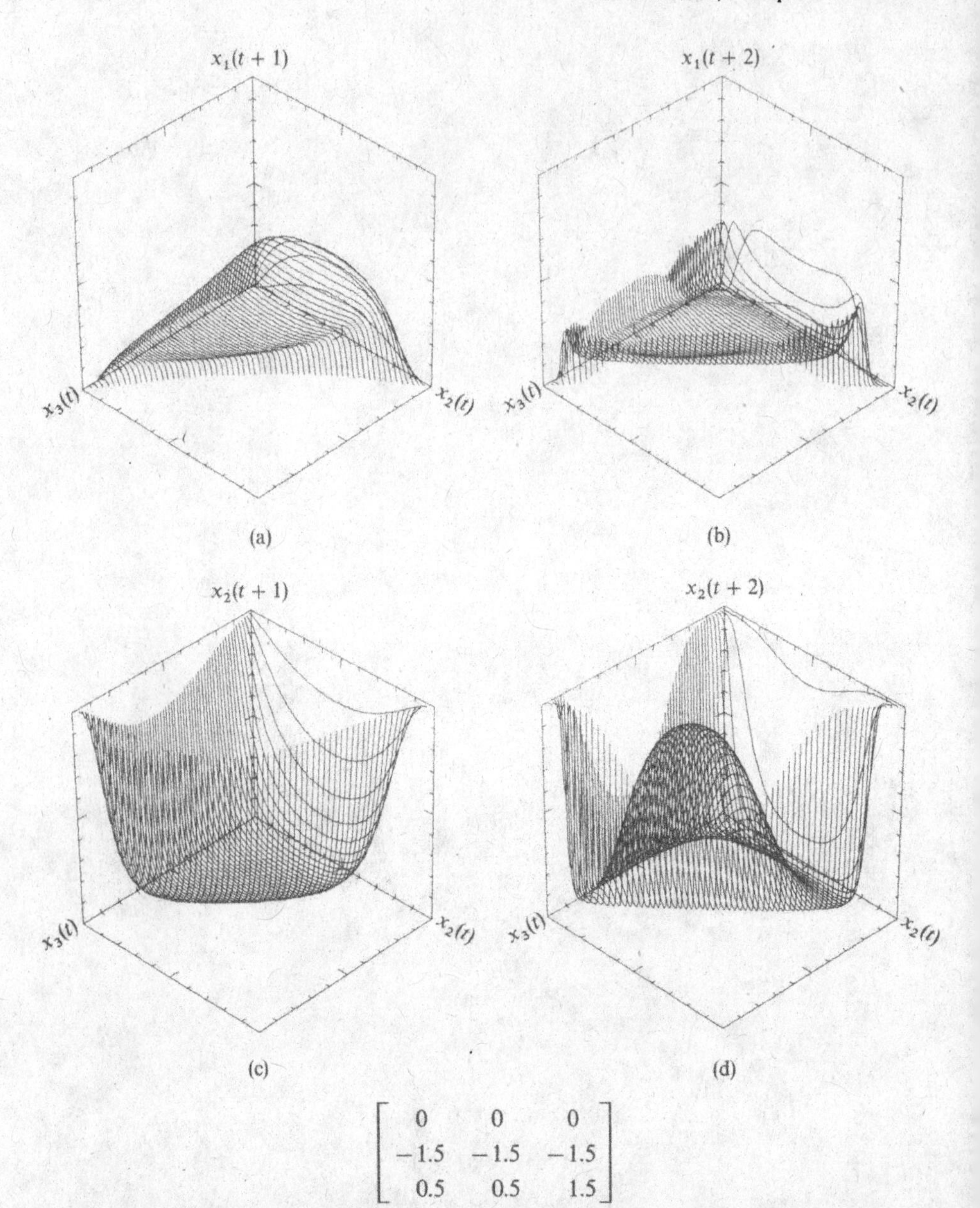

$$\begin{bmatrix} 0 & 0 & 0 \\ -1.5 & -1.5 & -1.5 \\ 0.5 & 0.5 & 1.5 \end{bmatrix}$$

FIGURE 32. The first and second iterates of x_1 and x_2 plotted against x_2 and x_3 for a stable four-period cycle: $A_1 = 1$, $A_2 = 10^{-3}$, $A_3 = 10^2$.

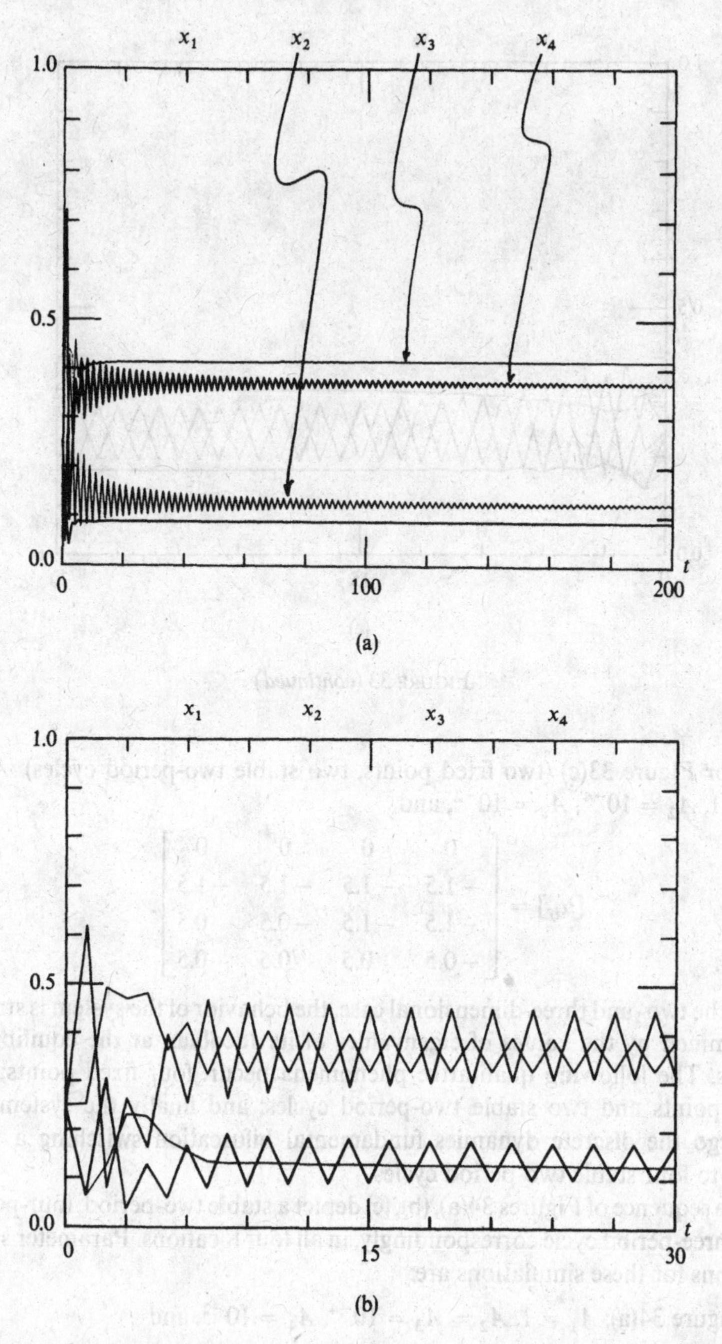

FIGURE 33. Stable behavior in the four-location, one-stock model: (a) four stable (fixed-point) equilibria; (b) one fixed point, and three stable two-period cycles; and (c) two fixed points and two stable two-period cycles. For parameter specifications, see text.

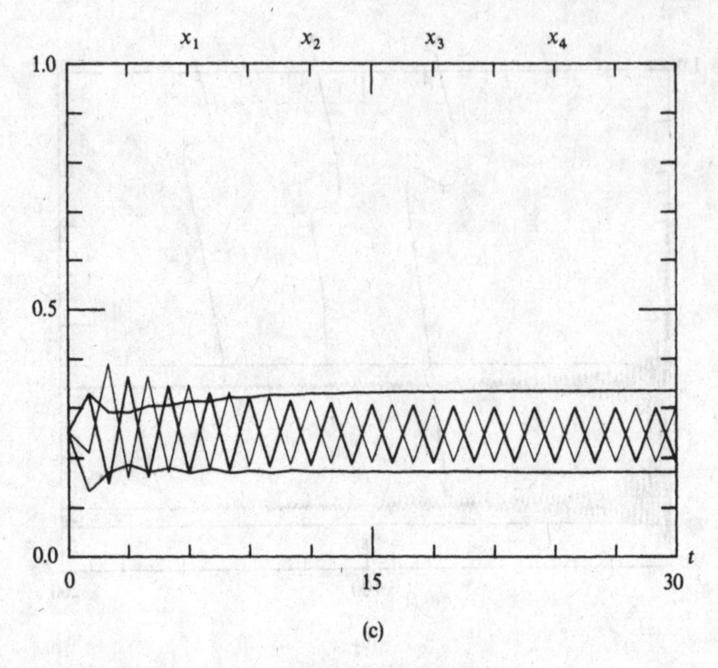

(c)

FIGURE 33 (*continued*)

and for Figure 33(c) (two fixed points, two stable two-period cycles): $A_1 = A_4 = 1$, $A_2 = 10^{-4}$, $A_3 = 10^{-2}$, and

$$[\alpha_{ij}] = \begin{bmatrix} 0 & 0 & 0 & 0 \\ -1.5 & -1.5 & -1.5 & -1.5 \\ -1.5 & -1.5 & -0.5 & 0.5 \\ -0.5 & 0.5 & 0.5 & -0.5 \end{bmatrix}.$$

As in the two- and three-dimensional case, the behavior of the system is strictly determined by the values of eigenvalues of its Jacobian at the equilibrium points. The following qualitative phenomena occur: four fixed points; two fixed points and two stable two-period cycles; and finally the system can undergo the discrete dynamics fundamental bifurcation switching a fixed point to four stable two-period cycles.

The sequence of Figures 34(a), (b), (c) depict a stable two-period, four-period and three-period cycle correspondingly, in all four locations. Parameter specifications for these simulations are:

for Figure 34(a): $A_1 = 1$, $A_2 = A_3 = 10^{-4}$, $A_4 = 10^{-2}$, and

$$[\alpha_{ij}] = \begin{bmatrix} 0 & 0 & 0 & 0 \\ -1.5 & -1.5 & -1.5 & -1.5 \\ -1.5 & -1.5 & -1.5 & -0.5 \\ -1.5 & 1.5 & -1.5 & -0.5 \end{bmatrix};$$

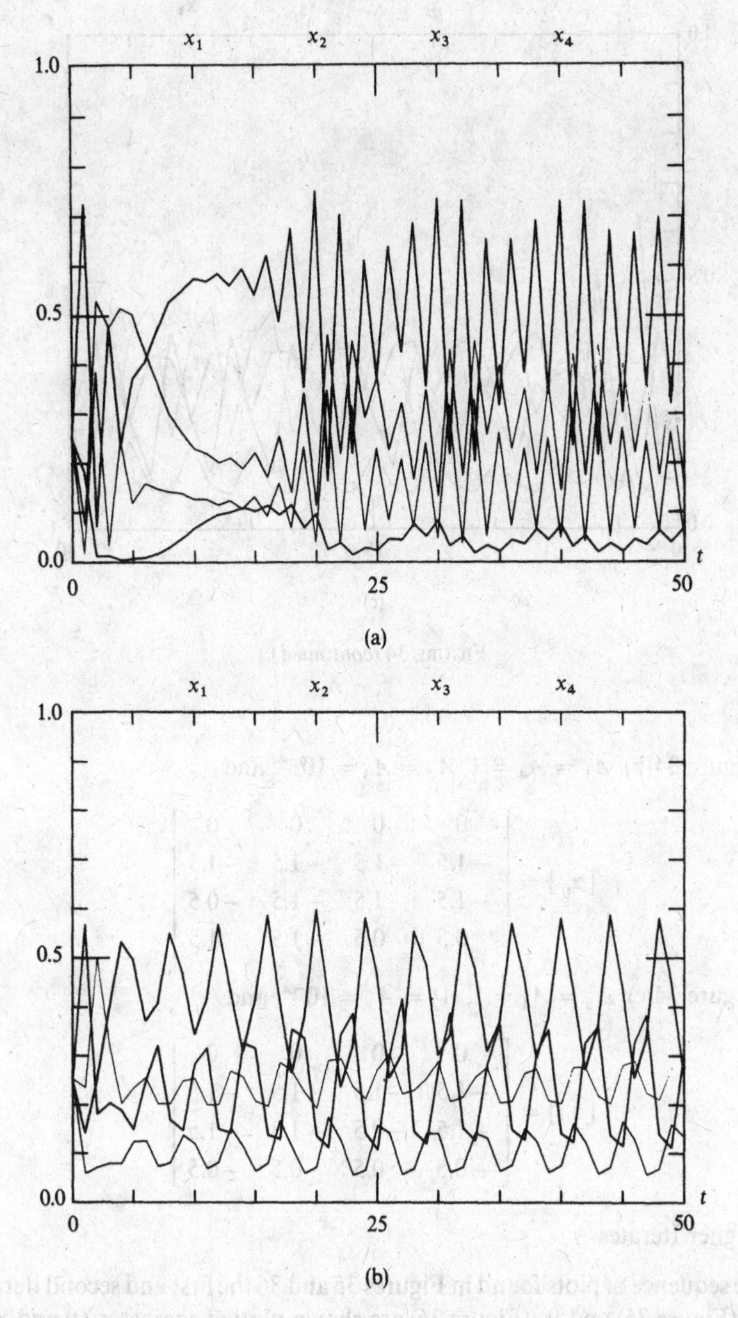

FIGURE 34. Regular periodic motion in the four-location, one-stock model: (a) a stable two-period cycle; (b) a stable four-period cycle; and (c) a stable three-period cycle. For parameters specifications, see text.

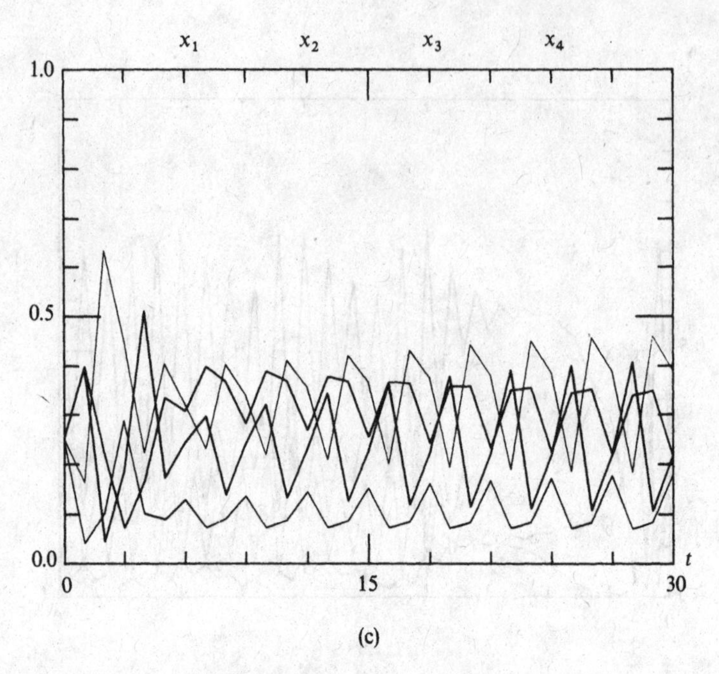

(c)

FIGURE 34 (*continued*)

for Figure 34(b): $A_1 = A_4 = 1$, $A_2 = A_3 = 10^{-4}$, and

$$[\alpha_{ij}] = \begin{bmatrix} 0 & 0 & 0 & 0 \\ -1.5 & -1.5 & -1.5 & -1.5 \\ -1.5 & -1.5 & -1.5 & -0.5 \\ 0.5 & 0.5 & -1.5 & 1.5 \end{bmatrix};$$

for Figure 34(c): $A_1 = A_4 = 1$, $A_2 = A_3 = 10^{-4}$, and

$$[\alpha_{ij}] = \begin{bmatrix} 0 & 0 & 0 & 0 \\ -1.5 & -1.5 & -1.5 & -1.5 \\ -1.5 & -0.5 & -1.5 & -1.5 \\ -0.5 & 0.5 & 0.5 & -0.5 \end{bmatrix}.$$

ii. Higher Iterates

In the sequence of plots found in Figures 35 and 36 the first and second iterates of x_1 (Figure 35) and x_2 (Figure 36) are shown plotted against $x_1(t)$ and $x_3(t)$ and for two different values of x_4 ($x_4(t) = 0.75$ and $x_4(t) = 0.50$): the case depicted identifies three stable two-period cycles and a fixed-point behavior in the four-location, one-stock case.

Model specifications for both Figures 35 and 36 are as follows: $A_1 = 1$,

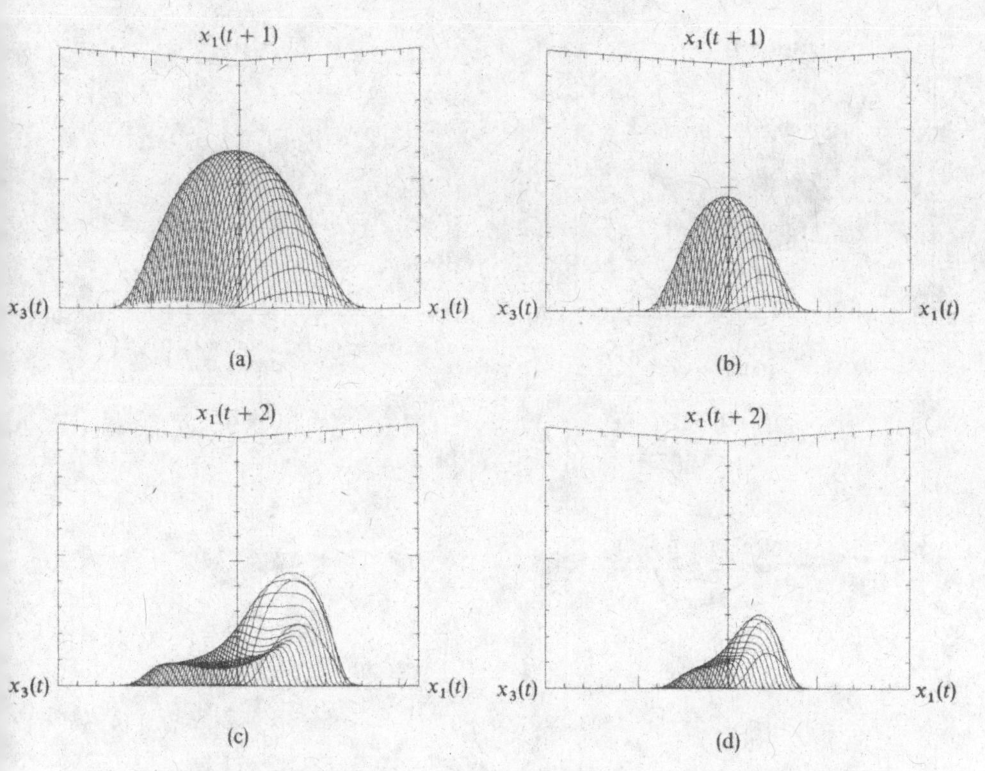

FIGURE 35. Higher iterates in the four-location, one-stock model. Three stable two-period cycles and one fixed-point behavior is shown for $x_1(t+1)$ and $x_1(t+2)$, and are plotted against $x_1(t)$ and $x_3(t)$ for $x_4(t) = 0.75$ (upper part) and $x_4(t) = 0.50$ (lower part). For parameter specifications, see text.

$A_2 = A_3 = A_4 = 10^{-4}$, and

$$[\alpha_{ij}] = \begin{bmatrix} 0 & 0 & 0 & 0 \\ -1.5 & -1.5 & -1.5 & -1.5 \\ -1.5 & -0.5 & -1.5 & -1.5 \\ -0.5 & -1.5 & -1.5 & -1.5 \end{bmatrix}.$$

An equivalent sequence is shown in Figures 37 and 38 for stable two-period cycles in all four locations.

Model specifications for both Figures 37 and 38 as follows: $A_1 = 1$, $A_2 = A_3 = 10^{-4}$, $A_4 = 10^{-2}$, and

$$[\alpha_{ij}] = \begin{bmatrix} 0 & 0 & 0 & 0 \\ -1.5 & -1.5 & -1.5 & -1.5 \\ -1.5 & -1.5 & -1.5 & -0.5 \\ -0.5 & 0.5 & -0.5 & -0.5 \end{bmatrix}.$$

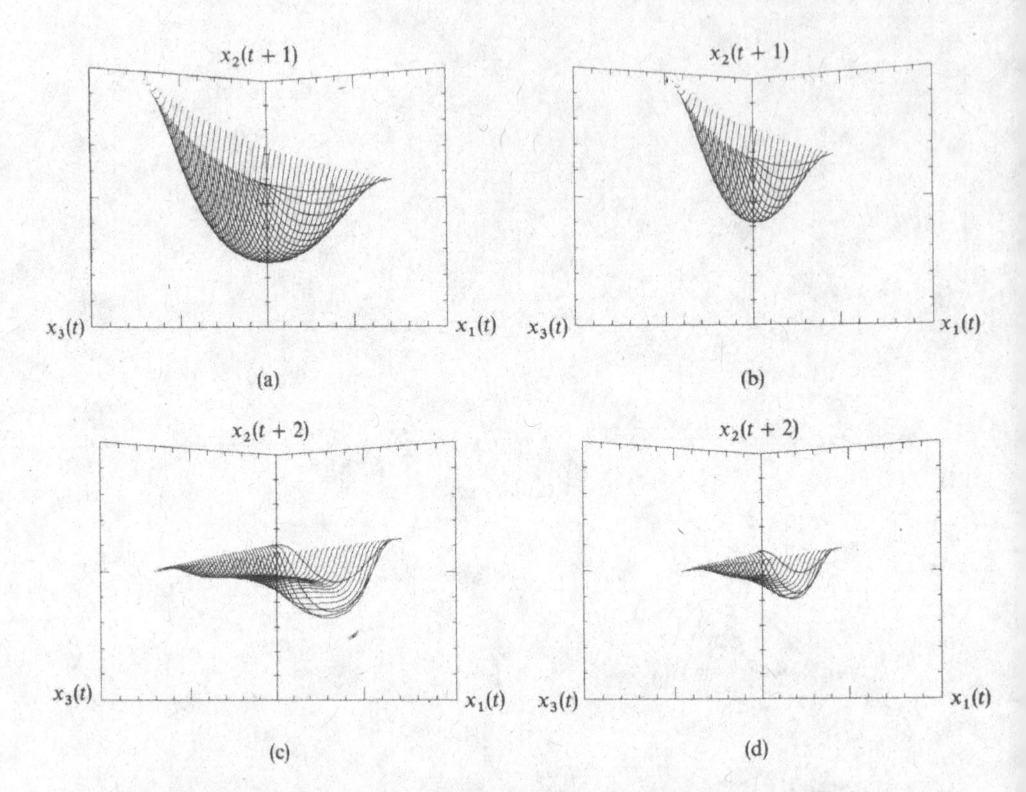

FIGURE 36. Iterates of x_2. They are plotted against $x_1(t)$ and $x_3(t)$ for $x_4(t) = 0.75$ (upper part) and $x_4(t) = 0.50$ (lower part) in the identical case with Figure 35.

A three-period cycle in all four locations is demonstrated in Figure 39. Model specifications for Figure 39 are as follows: $A_1 = A_4 = 1$, $A_2 = A_3 = 10^{-4}$, and

$$[\alpha_{ij}] = \begin{bmatrix} 0 & 0 & 0 & 0 \\ -1.5 & -1.5 & -1.5 & -1.5 \\ -1.5 & -0.5 & -1.5 & -1.5 \\ -0.5 & 0.5 & 0.5 & -0.5 \end{bmatrix}.$$

All Figures 35–39 show the three-dimensional projections of a four-dimensional surface as seen from the (x_1, x_3) plane. A view off that plane is shown for a case of chaotic motion in Figure 40. Cases where $x_4 < 0.5$ are not shown because of visual loss in detail due to scaling.

As in the simple May logistic model, iterative lagged behavior, $t + n$, where $n = 1, 2, \ldots, N$, depends on the slopes of the intersections of planes $x_i^*(t+n) = x_i^*(t)$, where $i = 1, 2, \ldots, I$, in this case (where $I = 4$) the equilibrium surface intersects four-dimensional surfaces $\underline{x}(t+n)$. Contrary to the May model, where the value $x_{\max}(t+n) = x_{\max}(t+1)$ and $x_{\min}(t+n) =$

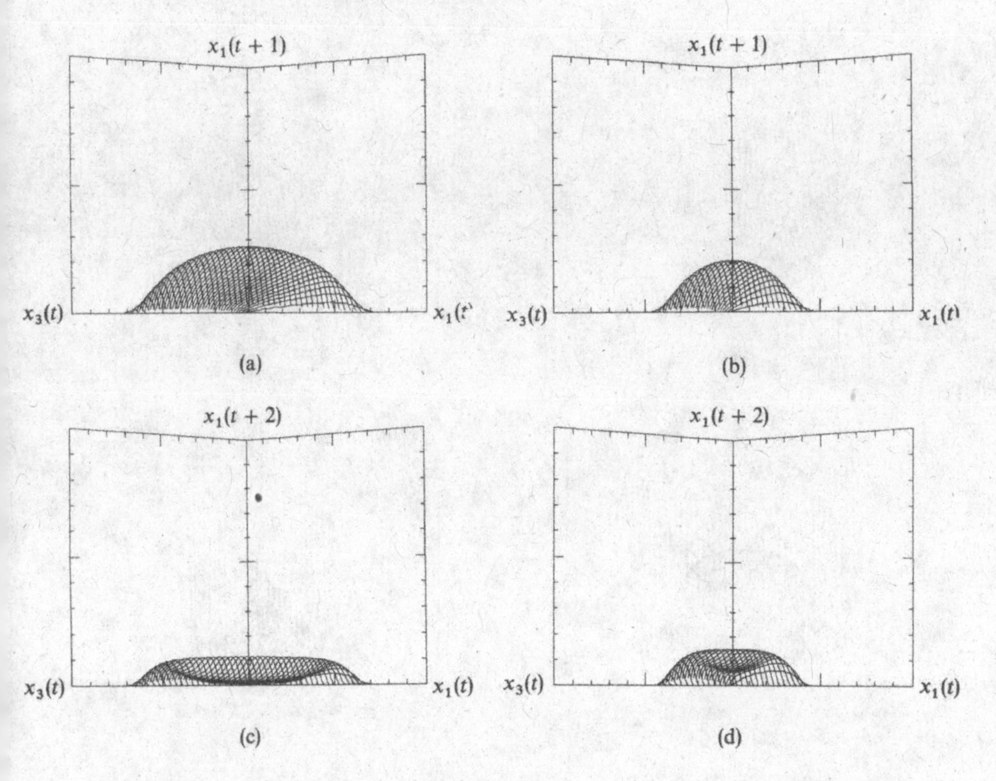

FIGURE 37. Iterates of x_1 for four, two-period cycles in the four-location, one-stock model. $x_1(t+1)$ and $x_1(t+2)$ are plotted against $x_1(t)$, $x_3(t)$ for $x_4(t) = 0.75$ (upper part) and $x_4(t) = 0.50$ (lower part). For model specifications, see text.

$x_{min}(t+2)$ for any $n = 1, 2, \ldots, N$, the higher iterates of the universal discrete relative dynamics algorithm do not necessarily remain at the same ranges established by the lower ($n = 1, 2$) iterates.

iii. Smooth Changes in Four Locations

An interesting phenomenon taking place in our discrete relative dynamics algorithm is the smoothness in its dynamics exhibited at specific ranges of parameter values. This special phenomenon observed in random runs of the four-dimensional case in the universal discrete relative dynamics algorithm allows for infinitesimal one-directional (or unimodal) movement along the state variables in the case of fixed-point behavior in all locations' stock. This phenomenon can be characterized as "quasi-continuous" dynamics in discrete-time algorithms.

The smooth movement is possible to last up to 2000 iterations, the maximum number of iterations allowed in these test runs, and it involves changes as small in magnitude as 10^{-9} in all locations' stock. It is not identified in random

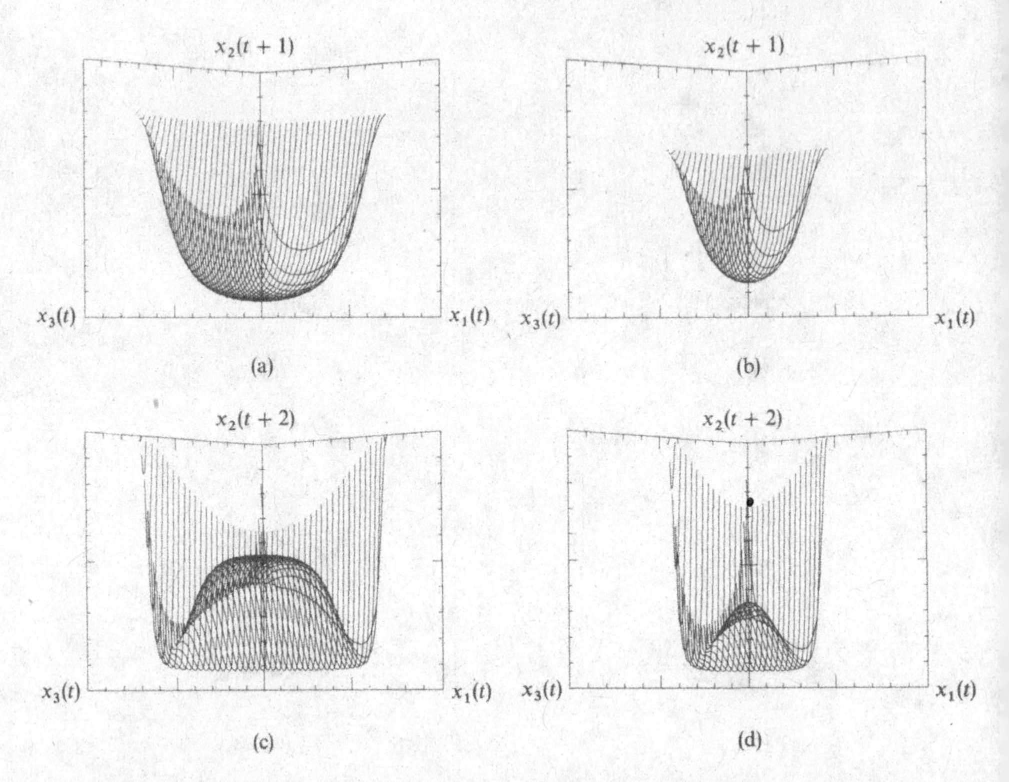

FIGURE 38. Iterates of x_2 for four, two-period cycles in the four-location, one-stock model. $x_2(t+1)$ and $x_2(t+2)$ are plotted against $x_1(t)$ and $x_3(t)$ for $x_4(t) = 0.75$ (upper part) and $x_4(t) = 0.50$ (lower part). Parameter specifications identical to those of Figure 37, see text.

runs for the two- and three-location, one-stock models without the computing system resulting in over- (or under-) flow. This event implies that as the dimensionality of the model increases the movement toward a noncompetitive exclusion equilibrium slows down significantly. Similar behavior is also depicted in the random runs of the two-location, two-stock model specifications.

c. HIGHER DIMENSIONS

i. Five Locations, and One Stock

Using random runs in the parameter (both $\underline{A}$ and $\underline{\alpha}$) space with a CYBER-205 system, the following results were obtained: generating random exponents between -1.5 and 1.5, and looping the A's from 10^{-10} to 10^{+10} with steps of 10^2, only one type of partial and local turbulence was produced. Usually, it involves a fixed point in one location, while the other four locations experience a stable two-period cycle. Turbulent regimes were also identified where various $n2^k$ cycles appear.

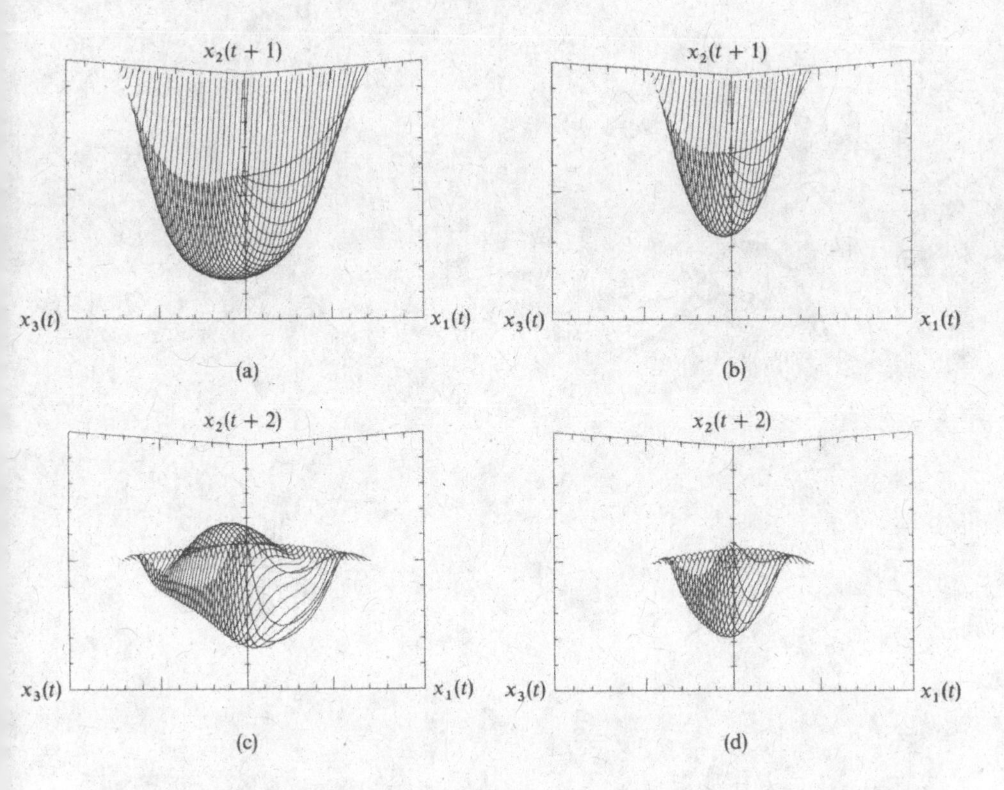

FIGURE 39. Iterates of x_2 for a stable three-period cycle in the four-location, one-stock model. $x_2(t+1)$ and $x_2(t+2)$ are plotted against $x_1(t)$ and $x_3(t)$ for $x_4(t) = 0.75$ (upper part) and $x_4(t) = 0.50$ (lower part). For parameter specifications, see text.

ii. Ten Locations, One Stock

Using random runs in the parameter ($\underline{A}$ and $\underline{\alpha}$) space on a VAX-11/750 system the following results were obtained: generating random exponents between -0.5 and 0.5 and looping the A's between 10^{-2} and 10^2 in steps of 10 no partial, local turbulence was detected. Fixed points, stable 2^k and $n2^k$ cycles, or chaos appear, together with unstable behavior. No slow movement in the state variables was obtained, as identified in cases involving the four-location, one-stock model.

As already mentioned in subsection 2 of Section B of this part, numerical simulations seem to indicate that the region of the parameter space where stable behavior occurs shrinks considerably around the core, when the dimensionality of the problem increases. This core, found at the vicinity of the point ($\underline{A} = 0$, $\underline{\alpha} = 0$) is a region of stability. Whether these results are the outcome of the simulation constraints of the machines used, or due to the dynamic properties of the algorithm proper is a subject of future research.

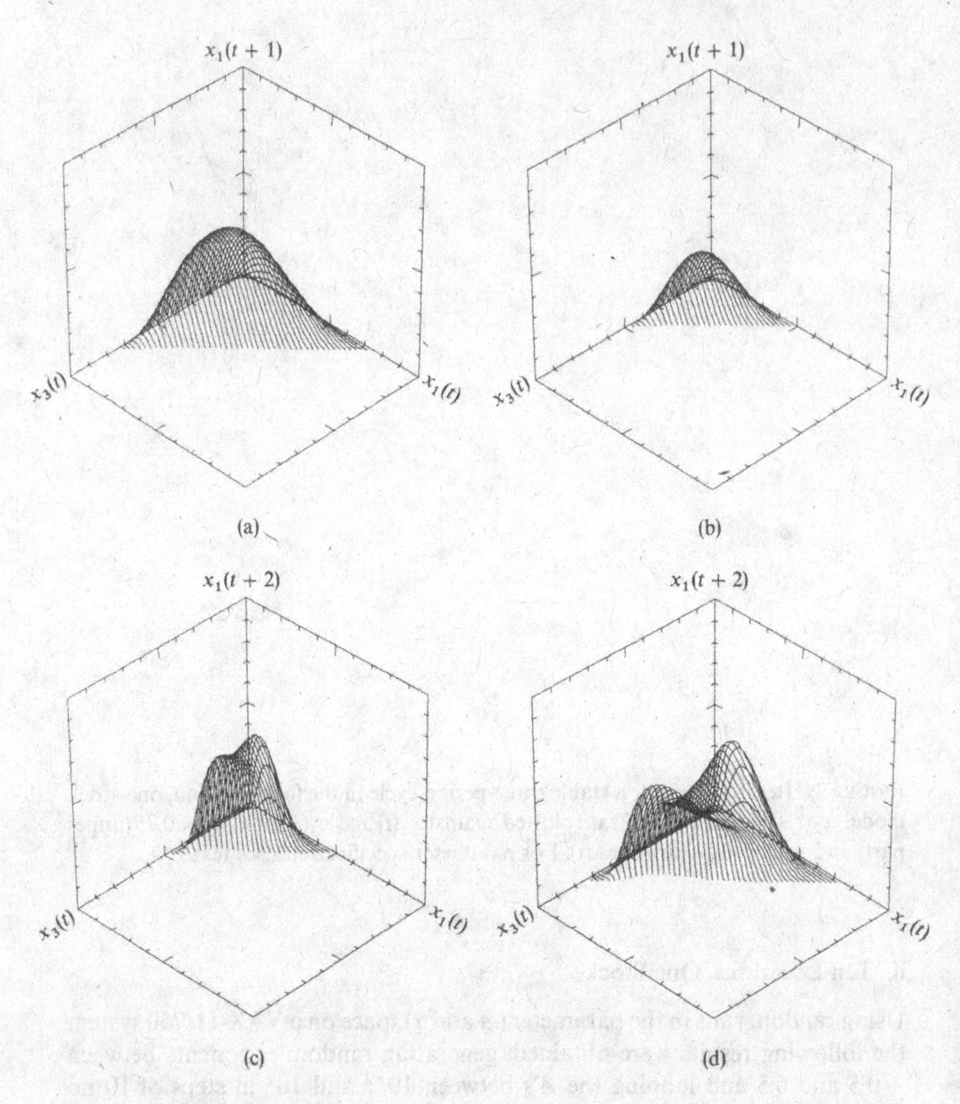

FIGURE 40. x_1 iterates in a four-location, one-stock case in the presence of chaos; a view off the (x_1, x_3) plane.

C. Empirical Evidence

1. *The Nine U.S. Regions and Their Aggregations*

In this subsection we present the only piece of empirical testing of the log-linear model available to date. It is based on a nine-region breakdown of the United States, following the Bureau of the Census spatial disaggregation. For a complete exposition of the geographical breakdown and associated issues of spatial composition, the reader is directed to Dendrinos and Sonis (1988). There, a full account of the findings is also provided.

A major advantage, along the efforts of statistically verifying dynamic models, which this particular discrete iterative dynamic process possesses over others (continuous or discrete maps) is its capacity to be tested using standard linear regression packages. Due to its log-linear form, one could easily employ any available statistical tests based on least-squares analysis.

Although the evidence is limited, it does seem to indicate statistical support for the discrete-time relative dynamics algorithm for the one-stock (population), three- and four-location model specifications. In carrying out these tests some more fundamental problems in empirical verification and theory construction in the social sciences emerged. They are dealt with in more detail in Dendrinos and Sonis (1988).

The U.S. Bureau of the Census aggregates the 50 states into nine areal divisions, which in turn are aggregated into four major regions. Our empirical testing utilizes this four-region breakdown to test the discrete-time algorithm. It also utilizes two more aggregations: a three-region breakdown and finally a two-region split of the U.S. space.

Data on population (the homogeneous stock), spatially distributed to these regions go back as far as 1850. This length in avaiable time series leads us to this particular set of tests with U.S. population data time series up to 1983. In total 133 years are involved with 63 counts. Projections for the year 2050 were also obtained, covering a 67-year time horizon.

The one-stock, I-location version of the discrete relative dynamics map under the log-linear specifications is

$$x_i(t+1) = \frac{F_i(t)}{\sum_j F_j(t)}; \qquad i, j = 1, 2, \ldots, I, \tag{III.C.1.1}$$

$$F_i(t) = A_i \prod_k x_k(t)^{a_{ik}} > 0; \qquad k = 1, 2, \ldots, I. \tag{III.C.1.2}$$

Dividing by the size of the (numéraire) region I, x_I, condition (III.C.1.1) becomes

$$y_i(t+1) = \sqrt{\frac{x_i(t+1)}{x_I(t+1)}} = \sqrt{\frac{F_i(t)}{F_I(t)}}, \tag{III.C.1.3}$$

and expressed in y's, it produces

$$y_i(t+1) = \frac{A_i}{A_I} \prod_k y_k(t)^{(a_{ik} - a_{Ik})}. \tag{III.C.1.4}$$

Transformed into logarithmic terms, a set of log-linear regression equations emerges

$$Y_i(t+1) = B_i + \sum_k \alpha_{ik} Y_k(t), \tag{III.C.1.5}$$

where

$$Y_i = \ln y_i \qquad \text{(III.C.1.6)}$$

$$B_i = \ln\left(\frac{A_i}{A_I}\right), \qquad \text{(III.C.1.7)}$$

$$\alpha_{ik} = a_{ik} - a_{Ik}. \qquad \text{(III.C.1.8)}$$

The system of equations in (III.C.1.5) can be used to test hypotheses on spatial stock dynamics, including population dynamics of the U.S. regions. It is, of course, of extreme interest to note that this system of equations can result in turbulence, for particular parameter values in the original system. The reason why this example was chosen as a testing ground for the map was that regional (state-wide) population counts in the United States are the most lengthy and consistent time series in socio-spatial sciences than any other stock series available.

2. *The Time Step and the Forces at Work*

Two time steps, i.e., real-time iteration sizes, were tested in the discrete-iterative process, a one-year step size and a ten-year one, in all runs involving one-stock, two-, three-, and four-location specifications. Some results are also discussed for a five-year step size.

Apparently, the broad (comprehensive, composite) forces at work at the various space–time disaggregations differ, depending on the space–time scale used to look at evolution. This was a major finding associated with the empirical testing, questioning radically the prevailing wisdom in social science modeling, that the form of the model must be independent of its disaggregation level (sectoral, or spatial, or temporal).

Long-term and large-scale forces (in this case the ten-year period iteration step size, at a two- or three-region spatial breakdown of the United States) compete with the short-term and small-scale forces (i.e., the one-year time step, and a four- (or more) region breakdown) in determining the evolution of each and every subregion of the United States.

Which one among these competing forces (and view-points) dominates the individual region's evolution may simply be a matter of location in space–time. Such dominance may not be stable in space–time, thus rendering the projections made by our map obsolete, for at least a subset of the regions, and over some of the time horizon under consideration.

3. *United States Regional Relative Population Instability*

Findings from the empirical tests (experiments?) indicate (Figure 41) that under a ten-year step and a four-region breakdown the Western region of the United States ends up dominating relative population distribution; whereas,

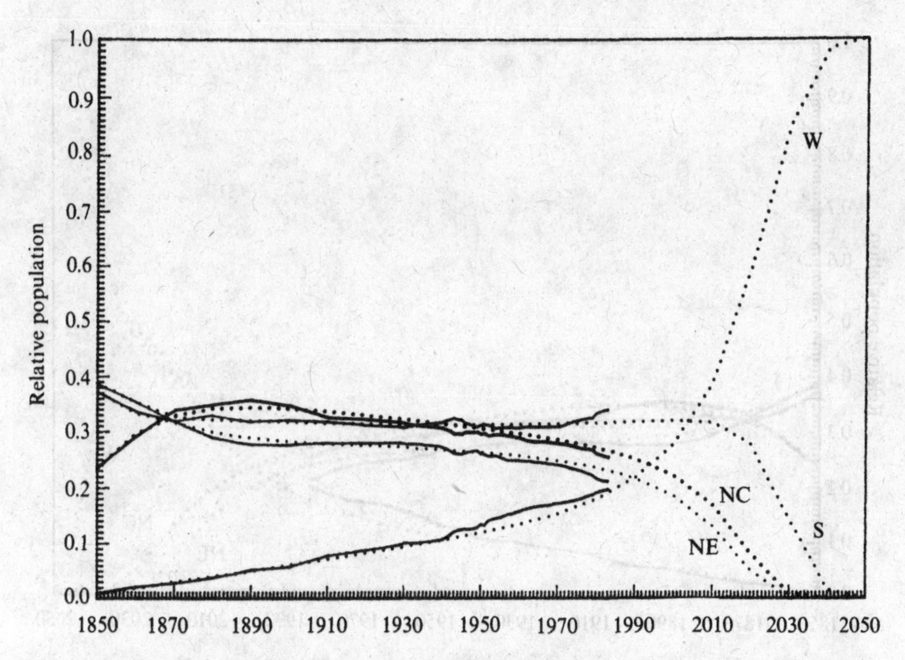

FIGURE 41. The four-region, ten-year period specifications of the United State's dynamic relative population distribution. Continuous lines show the observed paths, and dotted lines indicate simulation and projections. NE, North-East; NC, North-Central; S, South; and W, West. Reprinted by permission of the publisher from "Nonlinear Relative Discrete Population Dynamics of the U.S. Regions" by D.S. Dendrinos and M. Sonis, *Journal of Applied Mathematics and Computation*, 23:265–285, Copyright 1988 by Elsevier Science Publishing Co., Inc.

the other regions (North-East, North-Central, and South) gradually converge to competitive exclusion. Similarly, ambiguous findings emerge when a five-year step is tested for.

As an example the four-region, ten-year step size experiment from least squares stepwise regression results are shown below

$$x_{NE}(t+1) = \frac{1}{F(t)},$$

$$x_{NC}(t+1) = \frac{0.1432 x_{NE}(t)^{-0.9464} x_{NC}(t)^{0.345} x_S(t)^{-0.8312} x_W(t)^{-0.1238}}{F(t)},$$

$$x_S(t+1) = \frac{0.0024 x_{NE}(t)^{-1.877} x_{NC}(t)^{-0.8211} x_S(t)^{-1.9075} x_W(t)^{-0.2545}}{F(t)},$$

$$x_W(t+1) = \frac{0.0056 x_{NE}(t)^{-2.3661} x_{NC}(t)^{-1.272} x_S(t)^{-0.9506} x_W(t)^{-0.5795}}{F(t)},$$

where $F(t)$ is the sum of the four numerators of x_{NE}, x_{NC}, x_S, and x_W; and x_{NE}

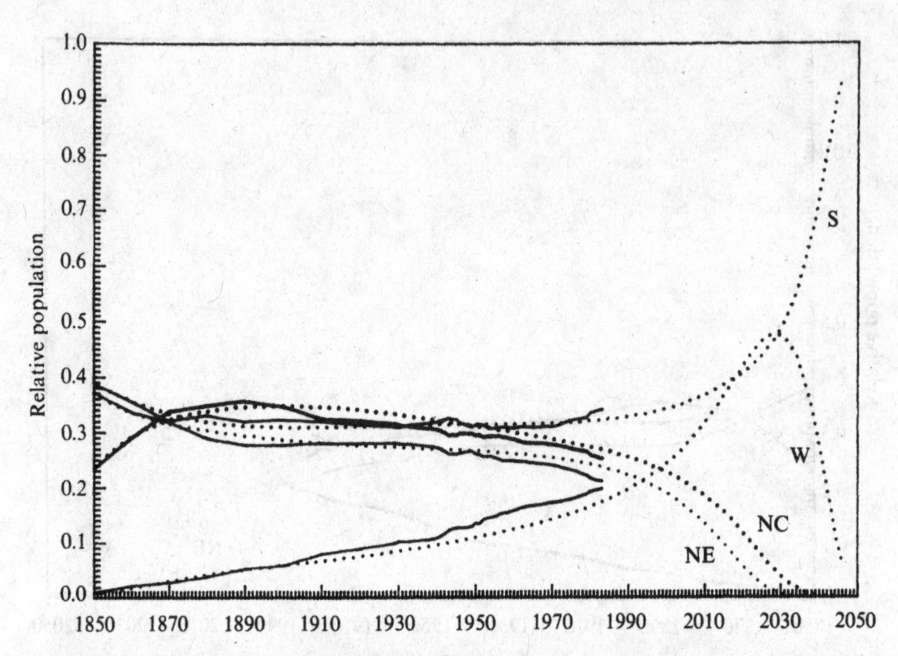

FIGURE 42. The four-region, one-year specifications of the United State's regional dynamic relative population distribution. Reprinted by permission of the publisher from "Nonlinear Relative Discrete Population Dynamics of the U.S. Regions" by D.S. Dendrinos and M. Sonis, *Journal of Applied Mathematics and Computation*, 23:265–285, Copyright 1988 by Elsevier Science Publishing Co., Inc.

is the relative population size of the North-East, x_{NC} is that of the North-Central, x_S of the Southern, and x_W of the Western United States' regions.

The statistical results seem to be robust in the immediate vicinity of these parameter values. However, relatively far away from these points in the parameter space the qualitative properties of these population dynamics may change, and possibly drastically. Such realization puts severe restraints on any attempt to extract a predictive (forecasting) value from any experiment in time series analysis, in this and in other cases, in the socio-spatial sciences. The extreme complexity involved in discrete or continuous dynamics, relative or absolute, and its implied constraints for statistical analysis is a major question with epistemological implications for the social sciences raised by nonlinear dynamical analysis. Some of these issues are raised by a number of researchers; see, e.g., the collection of papers in Casti (1984) and Barnett *et al.* (1988).

The case of a linear regression model is juxtaposed to the above findings. In a regional breakdown into three-locations *not involving regional population interdependencies*, a linear regression model using time as the only independent variable in linear form was tested. Results are shown in Figure 43. Naturally, this model violates the conservation condition which requires that all relative populations sum up to one at all time periods. Actually, this

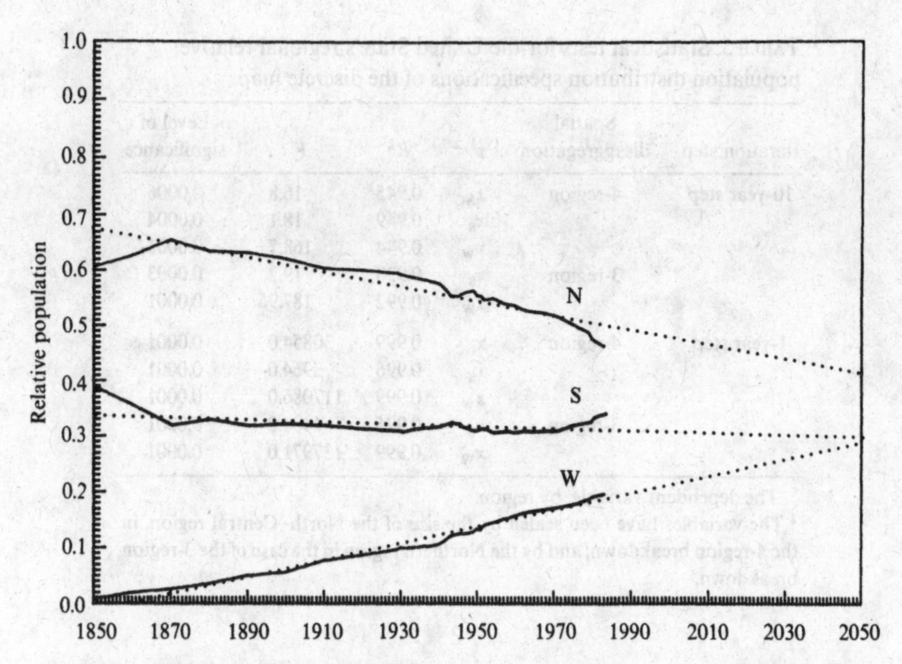

FIGURE 43. The three-region, no dynamic interdependencies, linear regression model of the United State's regional relative population distribution. Reprinted by permission of the publisher from "Nonlinear Relative Discrete Population Dynamics of the U.S. Regions" by D.S. Dendrinos and M. Sonis, *Journal of Applied Mathematics and Computation*, 23:265–285, Copyright 1988 by Elsevier Science Publishing Co., Inc.

condition is not violated more than 0.5% for the entire 1850–2050 period. Its goodness of fit is also excellent. The case is provided to make a number of points, in the next subsection.

4. *The Statistical Tests of Significance*

Working at the two-region breakdown, whereby the Western and Southern regions are merged into one superregion, the statistical results were not found to be admissible. Thus, for this particular spatial disaggregation of the problem, the log-linear model does not depict adequately the recorded regional comprehensive spatiotemporal population interdependencies, under any time-wise step size (one-, five-, or ten-year length.)

Results are, however, different at the three- and four-region breakdown. All cases, involving the one- and ten-year step size, were admissible, as was the linear model of Figure 43. In Table 5, the multiple R, the F-statistic, and corresponding levels of significance obtained for these runs are shown. Table 6 contains the results from the alternative linear time-dependent model not involving interdependencies. All these results, from both Tables 5 and 6, are within acceptable ranges.

TABLE 5. Statistical tests for the United State's regional relative population distribution specifications of the discrete map.

Iteration step	Spatial disaggregation	$x^{1,2}$	R^2	F	Level of significance
10-year step	4-region	x_{NC}	0.945	16.8	0.0006
		x_S	0.949	18.1	0.0004
		x_W	0.944	168.7	0.0001
	3-region	x_S	0.930	19.2	0.0003
		x_W	0.992	187.96	0.0001
1-year step	4-region	x_{NC}	0.999	30854.0	0.0001
		x_S	0.996	3964.0	0.0001
		x_W	0.999	117086.0	0.0001
	3-region	x_S	0.995	4404.0	0.0001
		x_W	0.999	137971.0	0.0001

[1] The dependent variable, by region.
[2] The variables have been scaled by the size of the North–Central region, in the 4-region breakdown; and by the Northern region in the case of the 3-region breakdown.

TABLE 6. Statistical tests of the linear, three-region model, not involving interdependencies.

Variable (relative regional size)	R^2	F	Level of significance
x_N	0.843	289.9	< 0.0001
x_S	0.434	12.5	0.0008
x_W	0.973	1970.0	< 0.0001

These findings bring about a central issue in validating social science hypotheses. All these models utilize a common data base. Population size of regions (composed of different states) in the United States is among the most reliable time series counts, in terms of accuracy, in all social science data available anywhere. Thus, findings emanating from these time series can be the least susceptible to quality and reliability questions.

The issue they bring about is the following: identical time series, at the same level of areal disaggregation, seem to support different hypotheses. For example, at the three-region breakdown two neighboring but different hypotheses (associated with the ten-year and one-year iteration steps, correspondingly) involving spatial and temporal interdependencies, and a nonneighboring one (associated with a linear, no-interdependencies model), cannot be rejected.

Both type of hypotheses have different implications for the United States regions under question. They imply different underlying forces at work during the calibration period (1850–1983), and a different future for the projection into time period (1983–2050).

D. Border Sequences in the One-Stock, Multiple-Location, Log-Linear Model

1. *Some General Results*

In this subsection we examine the domain in the state variable space where dynamic trajectories of the universal map are confined. The borders of these domains identify the limits of our strange containers. The strip containing state variable trajectories, for the case of log-linear comparative advantages producing F functions, will be denoted by two border sequences $[c_j^+(t), c_j^-(t)]$, such that

$$0 < c_j^-(t) \leq x_j(t) \leq c_j^+(t) < 1, \qquad j = 1, 2, \ldots, I. \qquad \text{(III.D.1.1)}$$

The above inequalities imply, for each $i = 1, 2, \ldots, I$, that

$$c_i^-(t) = 1 - \sum_{j \neq i} c_j^+(t) \leq x_i(t) \leq 1 - \sum_{j \neq i} c_j^-(t) = c_i^+(t). \qquad \text{(III.D.1.2)}$$

Border dynamic paths depend on the initial border $c_j^{\pm}(0)$ and the qualitative properties of the matrix $\|\text{sign } \alpha_{ij}\|$, where $[\alpha_{ij}]$ are the entries of the elasticity (structural parameter) matrix. If $\alpha_{ij} > 0$, then condition (III.D.1.1) implies

$$0 < c_j^-(t)^{\alpha_{ij}} \leq x_j(t)^{\alpha_{ij}} \leq c_j^+(t)^{\alpha_{ij}} < 1; \qquad \text{(III.D.1.3)}$$

if, on the other hand, $\alpha_{ij} < 0$, then (III.D.1.1) implies

$$1 \leq c_j^+(t)^{\alpha_{ij}} \leq x_j(t)^{\alpha_{ij}} \leq c_j^-(t)^{\alpha_{ij}} < \infty. \qquad \text{(III.D.1.4)}$$

As it is not practical to work for such broad ranges of the structural parameters, we will construct the border sequences for a special case where the border structure can be obtained analytically. We choose the case of

$$\begin{bmatrix} 0 & 0 & \cdots & 0 \\ \alpha_{21} & \alpha_{22} & \cdots & \alpha_{2I} \\ \vdots & \vdots & & \vdots \\ \alpha_{I1} & \alpha_{I2} & \cdots & \alpha_{II} \end{bmatrix},$$

where all $\alpha_{ij} \geq 0$. In this case, the inequalities of (III.D.1.3) imply

$$0 < A_i \prod_j c_j^-(t)^{\alpha_{ij}} \leq \frac{x_i(t+1)}{x_1(t+1)} = A_i \prod_j x_j(t)^{\alpha_{ij}} \leq A_i \prod_j c_j^+(t)^{\alpha_{ij}} < A_i. \qquad \text{(III.D.1.5)}$$

By summing, for $i \geq 2$, one obtains ($A_1 = 1$)

$$0 < \sum_{i \geq 2} A_i \prod_j c_j^-(t)^{\alpha_{ij}} \leq \frac{1 - x_1(t+1)}{x_1(t+1)} \leq \sum_{i \geq 2} A_i \prod_j c_j^+(t)^{\alpha_{ij}} < \sum_{i \geq 2} A_i.$$

which implies

$$\frac{1}{1 + \sum_{i \geq 2} A_i} < c_1^-(t+1) = \frac{1}{1 + \sum_{i \geq 2} A_i \prod_j c_j^+(t)^{\alpha_{ij}}} \leq x_1(t+1)$$

$$\leq \frac{1}{1 + \sum_{i \geq 2} A_i \prod_j c_j^-(t)^{\alpha_{ij}}} = c_1^+(t+1) < 1. \qquad \text{(III.D.1.6)}$$

For $i \geq 2$ one obtains from (III.D.1.6)

$$0 < c_i^-(t+1) = \frac{A_i \prod_j c_j^-(t)^{\alpha_{ij}}}{1 + \sum_{s\geq 2} \prod_j A_j c_j^+(t)^{\alpha_{ij}}} \leq x_i(t+1) = \frac{x_i(t+1)}{x_1(t+1)} x_1(t+1)$$

$$\leq \frac{A_i \prod_j c_j^+(t)^{\alpha_{ij}}}{1 + \sum_{s\geq 2} \prod_j A_j c_j^-(t)^{\alpha_{ij}}} = c_j^+(t+1) < A_i \qquad \text{(III.D.1.7)}$$

Thus, the sum of the environmental parameters set a lower bound for the movement in the state variables.

2. *Areas of State Variable Movement in the One-Stock, Three-Location, Log-Linear Model*

Strips which include the border sequences of state variable motion in the one-stock, three-location model of our universal map form a hexagon in the Möbius triangle. This hexagon provides only a very crude description of these domains, even in the case of a log-linear specification of the F functions.

Consider the matrix of elasticities

$$\begin{bmatrix} 0 & 0 & 0 \\ \alpha_{21} & \alpha_{22} & \alpha_{23} \\ \alpha_{31} & \alpha_{32} & \alpha_{33} \end{bmatrix},$$

so that

$$\frac{x_2(t+1)}{x_1(t+1)} = A_2 \prod_{j=1}^{3} x_j(t)^{\alpha_{2j}} = F_2,$$
$$\frac{x_3(t+1)}{x_1(t+1)} = A_3 \prod_{j=1}^{3} x_j(t)^{\alpha_{3j}} = F_3. \qquad \text{(III.D.2.1)}$$

From the above specifications one obtains

$$\frac{F_2^{\alpha_{31}}}{F_3^{\alpha_{21}}} = \frac{A_2^{\alpha_{31}}}{A_3^{\alpha_{21}}} x_2(t)^{-\begin{vmatrix}\alpha_{21} & \alpha_{22}\\ \alpha_{31} & \alpha_{32}\end{vmatrix}} x_3(t)^{-\begin{vmatrix}\alpha_{21} & \alpha_{23}\\ \alpha_{31} & \alpha_{33}\end{vmatrix}}. \qquad \text{(III.D.2.2)}$$

Let

$$\Delta_{12} = \begin{vmatrix}\alpha_{21} & \alpha_{22}\\ \alpha_{31} & \alpha_{32}\end{vmatrix}, \qquad \Delta_{13} = \begin{vmatrix}\alpha_{21} & \alpha_{23}\\ \alpha_{31} & \alpha_{33}\end{vmatrix}, \qquad \Delta_{23} = \begin{vmatrix}\alpha_{22} & \alpha_{23}\\ \alpha_{32} & \alpha_{33}\end{vmatrix},$$

and $A_2^{\alpha_{3i}}/A_3^{\alpha_{2i}} = B_i$, $i = 1, 2, 3$. Then

$$\frac{F_2^{\alpha_{31}}}{F_3^{\alpha_{21}}} = B_1 x_2(t)^{-\Delta_{12}} x_3(t)^{-\Delta_{13}}, \qquad \text{(III.D.2.3.i)}$$

$$\frac{F_2^{\alpha_{32}}}{F_3^{\alpha_{22}}} = B_2 x_1(t)^{\Delta_{12}} x_3(t)^{-\Delta_{23}}, \qquad \text{(III.D.2.3.ii)}$$

$$\frac{F_2^{\alpha_{33}}}{F_3^{\alpha_{23}}} = B_3 x_1(t)^{\Delta_{13}} x_2(t)^{\Delta_{23}}, \qquad \text{(III.D.2.3.iii)}$$

Let $0 < u, w < 1$; $0 < u + w \leq 1$. If $\alpha, \beta > 0$, then

$$0 < u^{\alpha} w^{\beta} \leq \left(\frac{\alpha}{\alpha+\beta}\right)^{\alpha} \left(\frac{\beta}{\alpha+\beta}\right)^{\beta}. \qquad \text{(III.D.2.4)}$$

If $\alpha, \beta < 0$, then

$$u^{\alpha} w^{\beta} \geq \left(\frac{\alpha}{\alpha+\beta}\right)^{\alpha} \left(\frac{\beta}{\alpha+\beta}\right)^{\beta}. \qquad \text{(III.D.2.5)}$$

Both conditions (III.D.2.4, 5) are the result of the theory of elementary inequalities. They directly imply, from (III.D.2.3.i–iii), that (if $\Delta_{12}, \Delta_{13}, \Delta_{23} \geq 0$)

$$\frac{F_2^{\alpha_{31}}}{F_3^{\alpha_{21}}} \geq B_1 \left(\frac{\Delta_{12}}{\Delta_{12}+\Delta_{13}}\right)^{-\Delta_{12}} \left(\frac{\Delta_{13}}{\Delta_{12}+\Delta_{13}}\right)^{-\Delta_{13}} = C_1, \qquad \text{(III.D.2.6)}$$

$$\frac{F_2^{\alpha_{33}}}{F_3^{\alpha_{23}}} \leq B_3 \left(\frac{\Delta_{13}}{\Delta_{13}+\Delta_{23}}\right)^{\Delta_{13}} \left(\frac{\Delta_{23}}{\Delta_{13}+\Delta_{23}}\right)^{\Delta_{23}} = C_3. \qquad \text{(III.D.2.7)}$$

The above imply that the domain of trajectories for the three state variables is given by the area in the Möbius triangle between curves

$$B_1 x_1(t+1)^{\alpha_{31}-\alpha_{21}} = \frac{x_2(t+1)^{\alpha_{33}}}{x_3(t+1)^{\alpha_{21}}}, \qquad \text{(III.D.2.8)}$$

$$\frac{x_2(t+1)^{\alpha_{31}}}{x_3(t+1)^{\alpha_{21}}} = C_1 x_1(t+1)^{\alpha_{31}-\alpha_{21}}, \qquad \text{(III.D.2.9)}$$

$$\frac{x_2(t+1)^{\alpha_{33}}}{x_3(t+1)^{\alpha_{23}}} = C_3 x_1(t+1)^{\alpha_{33}-\alpha_{23}}. \qquad \text{(III.D.2.10)}$$

Consider the special case of this log-linear model where $A_1 = A_2 = A_3 = 1$, and the matrix of the structural parameters is

$$\begin{bmatrix} 0 & 0 & 0 \\ 1 & 1 & 1 \\ 2 & 0 & 0 \end{bmatrix}.$$

Now

$$\Delta_{12} = \Delta_{13} = \begin{vmatrix} 1 & 1 \\ 2 & 0 \end{vmatrix} = -2; \quad \Delta_{23} = \begin{vmatrix} 1 & 1 \\ 0 & 0 \end{vmatrix} = 0, \qquad \text{(III.D.2.11)}$$

and

$$\frac{F_2^2}{F_3} = x_2(t)^2 x_3(t)^2; \quad \frac{1}{F_3} = x_1(t)^{-2}, \qquad \text{(III.D.2.12)}$$

or

$$\frac{1}{16} \geq \left[\frac{x_2(t+1)}{x_1(t+1)}\right]^2 \bigg/ \left[\frac{x_3(t+1)}{x_1(t+1)}\right]; \quad 1 \leq \frac{x_1(t+1)}{x_3(t+1)}, \qquad \text{(III.D.2.13)}$$

or

$$\begin{cases} \frac{1}{16} x_1(t+1) x_3(t+1) \geq x_2(t+1)^2, \\ x_3(t+1) \leq x_1(t+1). \end{cases} \qquad \text{(III.D.2.14)}$$

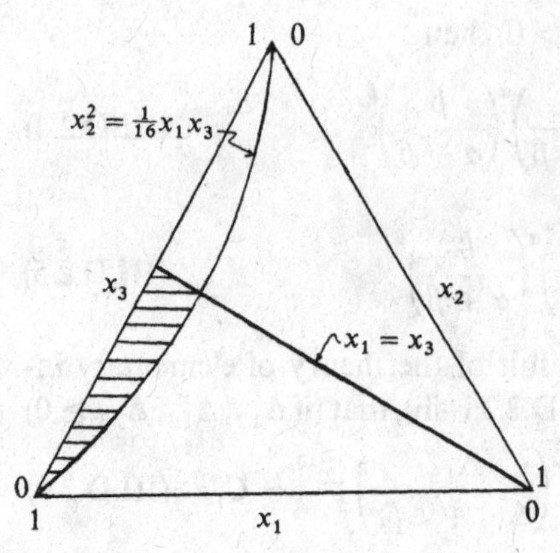

FIGURE 44. The boundaries of a strange container. A log-linear model with $A_1 = A_2 = A_3 = 1$ and the matrix of elasticities $\begin{bmatrix} 0 & 0 & 0 \\ 1 & 1 & 1 \\ 2 & 0 & 0 \end{bmatrix}$.

The geometric configuration of this strange container is the parabolic curve, $x_2(t+1)^2 = \frac{1}{16}x_1(t+1)x_3(t+1)]$, and the straight line $x_1(t+1) = x_3(t+1)$ (Figure 44).

The simplest case for a strange container to form is when the matrix of elasticities contains proportional row entries: $\alpha_{31}/\alpha_{21} = \alpha_{32}/\alpha_{22} = \alpha_{33}/\alpha_{23} = k$. In this case the determinants $\Delta_{12} = \Delta_{13} = \Delta_{23} = 0$ and conditions (III.D.2.3.i–iii) result in

$$A_3 x_2(t+1)^k = A_2^k x_1(t+1)^{k-1} x_3(t+1), \qquad \text{(III.D.2.15)}$$

which gives the curve generating the strange container.

E. One-Stock, Multiple-Location, Discrete-Time, Logistic Growth

1. *Definition and Central Analytical Properties*

Here we present a variance of our universal map to indicate the discrete-time, discrete-space equivalent of the continuous time logistic model. We further link this submodel to the multinomial logistic growth case.

Discrete-time logistic growth is given by the following specifications of the one-species, I-location model

$$\begin{aligned} x_i(t+1) &= \frac{x_i(t)F_i(t)}{\sum_{j=1}^{I} x_j(t)F_j(t)}, \\ 0 &\le x_i(t) \le 1, \qquad i = 1, 2, \ldots, I, \\ \sum_{i=1}^{I} x_i(t) &= 1, \\ 0 &< F_i(t) = F_i(x_1(t), x_2(t), \ldots, x_I(t)). \end{aligned} \qquad \text{(III.E.1.1)}$$

Any increment in species relative abundance at any location i, $[x_i(t+1) -$

$x_i(t)]$, is given by

$$x_i(t+1) - x_i(t) = x_i(t+1) \sum_{j=1}^{I} x_j(t)\left[1 - \frac{F_j(t)}{F_i(t)}\right],$$

and, consequently, the rate of change in relative abundance is provided by

$$\frac{x_i(t+1) - x_i(t)}{x_i(t+1)} = \sum_{j=1}^{I} x_j(t)\left[1 - \frac{F_j(t)}{F_i(t)}\right], \quad i = 1, 2, \ldots, I. \qquad \text{(III.E.1.2)}$$

This system of difference equations represents a discrete-time analogue of the continuous-time logistic differential equations model

$$\frac{d \ln x_i(t)}{dt} = \sum_{j=1}^{I} x_j(t) a_{ij}(t), \qquad i = 1, 2, \ldots, I. \qquad \text{(III.E.1.3)}$$

Considering that for any pair of locations the following condition holds cumulatively over a time horizon t

$$\frac{x_i(t)}{x_s(t)} = \frac{x_i(0)}{x_s(0)} \prod_{\tau=0}^{t-1} \frac{F_i(\tau)}{F_s(\tau)}, \qquad \text{(III.E.1.4)}$$

where $x_i(0)$, $x_s(0)$ are the initial perturbations at the two locations, one obtains the generalized logistic growth model

$$x_i(t) = \frac{x_i(0) \prod_{\tau=0}^{t-1} F_i(\tau)}{\sum_{s=1}^{I} x_s(0) \prod_{\tau=0}^{t-1} F_s(\tau)}, \qquad i = 1, 2, \ldots, I. \qquad \text{(III.E.1.5)}$$

In the particular case of constant functions $F_i = e^{a_i}$, $i = 1, 2, \ldots, I$, the standard, multinomial, logistic growth model is obtained (Sonis, 1987),

$$x_i(t) = \frac{x_i(0) e^{a_i t}}{\sum_{s=1}^{I} x_s(0) e^{a_s t}}, \qquad i = 1, 2, \ldots, I. \qquad \text{(III.E.1.6)}$$

2. *The Jacobi Matrix*

The Jacobi slope matrix J, see (III.A.1.17), has the following components, for $i \neq j$ cross-locational effects,

$$s_{ij}(t+1, t) = \frac{\partial x_i(t+1)}{\partial x_j(t)}$$

$$= x_i(t+1) \sum_{\substack{s=1 \\ s \neq j}}^{I} x_s(t+1) \frac{\partial}{\partial x_j(t)} \ln \frac{F_i(t)}{F_s(t)} - \frac{x_i(t+1) F_j(t)}{\sum_{s=1}^{I} x_s(t) F_s(t)}. \qquad \text{(III.E.2.1)}$$

For $i = j$

$$s_{ii}(t+1, t) = \frac{\partial x_i(t+1)}{\partial x_i(t)}$$

$$= x_i(t+1) \sum_{\substack{s=1 \\ s \neq i}}^{I} x_s(t+1) \frac{\partial}{\partial x_i(t)} \ln \frac{F_i(t)}{F_s(t)} + \frac{1 - x_i(t+1) F_i(t)}{\sum_{s=1}^{I} x_s(t) F_s(t)}. \qquad \text{(III.E.2.2)}$$

At the equilibrium point $(x_1^*, x_2^*, \ldots, x_I^*)$ the J^* matrix has entries

$$s_{ij}^* = x_i^* \sum_{\substack{s=1 \\ s \neq j}}^{I} x_s^* \left[\frac{\partial}{\partial x_j(t)} \ln \frac{F_i(t)}{F_s(t)} \right]^* - \frac{x_i^* F_j^*}{\sum_{s=1}^{I} x_s^* F_s^*}, \qquad i \neq j,$$
$$s_{ii}^* = x_i^* \sum_{\substack{s=1 \\ s \neq i}}^{I} x_s^* \left[\frac{\partial}{\partial x_i(t)} \ln \frac{F_i(t)}{F_s(t)} \right]^* + \frac{(1 - x_i^*) F_i^*}{\sum_{s=1}^{I} x_s^* F_s^*}. \qquad \text{(III.E.2.3)}$$

3. *Equilibria*

Equilibrium states satisfy the conditions

$$x_i^* = \frac{x_i^* F_i^*}{\sum_{j=1}^{I} x_j^* F_j^*}, \qquad i = 1, 2, \ldots, I, \qquad \text{(III.E.3.1)}$$

implying immediately that the competitive exclusion states ($x_r^* = 1, x_i^* = 0, i \neq r, i = 1, 2, \ldots, I$) *are equilibrium states of the generalized logistic growth version of our universal map.*

It can be shown (see Sonis, 1989), that partial competitive exclusion equilibria of the type

$$\begin{aligned} & x_s^* \neq 0, \qquad s = 1, 2, \ldots, S \quad (i \leq S \leq I), \\ & x_i^* = 0, \qquad i \neq s, \quad i = S, \; S+1, \ldots, I, \\ & \sum_{s=1}^{S} x_s^* = 1, \end{aligned} \qquad \text{(III.E.3.2)}$$

where by s we have designated those locations with some species abundance, can be derived but are in general unstable. A special case, that of the standard, multinomial, logistic growth, seems to indicate that partial competitive exclusion equilibria are stable, but not asyptotically so.

Conclusions

A number of innovative phenomena were uncovered in the one-stock, multiple-location, discrete-time relative dynamics. Central among them was the discovery of "local" and "partial" turbulence. It was classified as "local" because it is possible to confine it to a subset of the competing regions for the (homogeneous) stock. It was termed "partial" because it only involves stable periodic motions coexisting with fixed-point behavior, but not evolving into a full cascade of bifurcations leading to chaos. Whether higher-period cycles can also be found, coexisting with a fixed attractor and possibly generated by the splitting of the two-period cycle or the fixed point, still remains unknown.

Alongside "strange attractors," areas in the state variable space where orbits gravitate toward each other from either the inside or the outside, we also found "strange containers" of orbits. These are areas in the state variable space where orbits are limited, never exiting these regions no matter what the initial perturbation.

Strange containers were found to be either highly dense (localized) in the state variable space, or they could be thin and global, spanning almost the whole admissible domain. These phenomena, found scattered in the parameter space, were supplemented with hybrid "quasi-container–attractor" cases of orbital motion.

May-type chaos, obeying Feigenbaum's constants, over the parameter or slope sequences, were found to be special cases. In general, there are apparently different types of chae with varying sequences of events leading to them. Li and Yorke's condition, that "period-three implies chaos," does not seem to be a statement applicable to the universal algorithm of discrete-time relative dynamics, to the extent that it must immediately precede chaotic movement and follow all even cycles in a period-doubling cascade as well as all odd-period cycles.

Limitations were discussed on the numerical simulation procedures used to explore the behavior of the state variables in the various regions of the parameter space. As the problem's dimensionality increases the region of the parameter space, allowed to record stable dynamic behavior, shrinks fast.

As well as efforts toward finding new phenomena embedded in the parameter space of the universal map, attention must be focused on how to improve the simulating capacity of numerical methods. Clearly, some of the instability portrayed must be due to the numerical simulation limitations and the ability of the computing machine used to approximate, rather than precisely identify, the properties of the universal map.

Some criteria must also be devised to search in the vast regions of the universal map's parameter space for innovative dynamic events in the state variable space.

We supplied limited empirical evidence regarding the United States relative population distribution into various regions. Although the evidence was not extensive, support for the social-science-related applicability of the algorithm seems to have emerged out of the statistical tests. At the level of areal disaggregation considered, the spatial relative United States regional population dynamics seem to be unstable when looked at within a 200-year period (1850–1983 of recorded time series, and 1983–2050 in forecast.) This is the longest time period, to date, used by any model in the social science literature to calibrate and test a particular hypothesis. Results of these simulations indicate that typically, the northern regions of the United States converge in the long term to abandonment.

Competitive exclusion seems to characterize the dynamic relative population size distribution of the United States regions. This type of dynamic equilibria are the only configurations obtained by our variation of the universal map producing a discrete-time equivalent of the continuous-time logistic growth model. The latter, it was demonstrated, can produce partial competitive exclusion, as well, under particular conditions, like for intance those associated with the standard multinomial logit model.

Next, we proceed to present the multiple-stock, multiple-location specification of our universal map.

Part IV
Multiple Stocks, Multiple Locations

Summary

In this part we briefly explore the general J-species, I-location model of our universal algorithm by providing some general formulas and conditions surrounding the equilibria configurations. Certain numerical simulations supplement the analysis, which becomes, due to its dimensionality, largely intractable. The phenomena identified in earlier versions of our map are repeated, although the region of fixed-point behavior (coexistence-type equilibria for all stocks at all locations) in the parameter space has shrunk considerably.

We start with the general conditions found in the J-stock, I-location specifications of a general set of comparative advantage generating functions, and gradually narrow our analysis to log-linear specifications of these functions and to the two-stock case. Some results for a discrete-time equivalent of continuous-time logistic growth dynamics is also presented, with emphasis placed on the two-stock, multiple-location model.

A. The General Model

1. *Analytical Results*

We now assume that there are J different stocks, possibly located at I different (heterogeneous) locations. Each stock j ($j = 1, 2, \ldots, J$) is homogeneous. Examples of such stocks (intragroup homogeneous, intergroup heterogeneous) could be: J distinct population (or labor) types; J different built capital stocks (for instance, classified according to vintage) or J stocks of financial capital (currencies); J types of economic output (product types); wealth; or any other economic, social, political, and other types of variable, or a combination of them.

Interest, as always in relative dynamics, is not on the total (absolute) level of these stocks in space–time; but rather on their relative distribution in space–time. There are a number of possible outcomes one might expect in reference to these allocations. Extinction of one stock at one location; abandonment of one location by all stocks; extinction of one stock from all but one location; and allocation of certain stocks only on certain locations.

The general case will be presented in the following form of discrete-time, relative dynamics

$$
\begin{aligned}
&x_{ji}(t+1) = \frac{F_{ji}(x_{ji}(t); j = 1, 2, \ldots, J; i = 1, 2, \ldots, I)}{\sum_{s=1}^{I} F_{js}},\\
&j = 1, 2, \ldots, J, \quad i = 1, 2, \ldots, I,\\
&\sum_{i=1}^{I} x_{ji}(t+1) = 1; \qquad j = 1, 2, \ldots, J,\\
&F_{ji} > 0,\\
&0 < x_{ji} < 1; \qquad j = 1, 2, \ldots, J; \quad i = 1, 2, \ldots, I,\\
&0 < x_{ji}(0) < 1; \qquad \sum_{i=1}^{I} x_{ji}(0) = 1; \qquad j = 1, 2, \ldots, J.
\end{aligned}
\tag{IV.A.1.1}
$$

Assume a numéraire location (I); then

$$\frac{x_{ji}(t+1)}{x_{jI}(t+1)} = \frac{F_{ji}}{F_{jI}}, \qquad i = 1, 2, \ldots, I; \quad j = 1, 2, \ldots, J. \tag{IV.A.1.2}$$

Cross-location, cross-stock effects are given by the conditions

$$s_{qp}^{ji}(t+1, t) = \frac{\partial x_{ji}(t+1)}{\partial x_{qp}(t)} = x_{ji}(t+1) \sum_{\substack{s=1\\ s\neq 1}}^{I} x_{js}(t+1) \frac{\partial}{\partial x_{qp}(t)} \ln \frac{F_{ji}}{F_{js}}. \tag{IV.A.1.3}$$

Since $\sum_i x_{ji}(t+1) = 1$, from (IV.A.1.3) it follows that

$$\sum_i s_{qp}^{ji}(t+1, t) = \sum_i \frac{\partial x_{ji}(t+1)}{\partial x_{qp}(t)} = 0, \qquad j, q = 1, 2, \ldots, J; \quad p = 1, 2, \ldots, I. \tag{IV.A.1.4}$$

In a complete form, the cross-location, cross-stock effects can be presented by the Jacobi block-matrix

$$J(t+1, t) = \|J_{pi}(t+1, t)\|, \qquad p, i = 1, 2, \ldots, I, \tag{IV.A.1.5}$$

where each block is represented by matrices

$$J_{pi}(t+1, t) = \|s_{qp}^{ji}(t+1, t)\| = \left\| \frac{\partial x_{ji}(t+1)}{\partial x_{qp}(t)} \right\|, \qquad q, j = 1, 2, \ldots, J. \tag{IV.A.1.6}$$

Due to conditions (IV.A.1.4) the Jacobian det $J(t+1, t) = 0$ and moreover, the rank of the Jacobi block-matrix $J(t+1, t)$ is equal to $[IJ - (I + J)]$.

The equilibrium state of the discrete-time, discrete-space, relative dynamics model containing I locations and J stocks has components x_{ji}^* given by

$$x_{ji}^* = 1/1 + \frac{1}{F_{ji}^*} \sum_{\substack{s=1 \\ s \neq i}}^{I} F_{js}^*, \qquad i, s = 1, 2, \ldots, I, \quad j = 1, 2, \ldots, J, \quad \text{(IV.A.1.7)}$$

or, put differently,

$$\frac{x_{ji}^*}{x_{jk}^*} = \frac{F_{ji}^*}{F_{jk}^*}, \qquad i, k = 1, 2, \ldots, I; \quad j = 1, 2, \ldots, J. \qquad \text{(IV.A.1.8)}$$

At equilibrium, the Jacobi block-matrix $J^* = \|J_{pi}^*\|$ includes blocks of the form

$$J_{pi}^* = \|s_{qp}^{*ji}\| = \left\| x_{ji}^* \sum_{s=1}^{I} x_{js}^* \left[\frac{\partial}{\partial x_{qp}(t)} \ln \frac{F_{ji}(t)}{F_{js}(t)} \right]^* \right\|. \qquad \text{(IV.A.1.9)}$$

If all eigenvalues λ of the Jacobi block-matrix J^* satisfy the condition $|\lambda| < 1$, then the equilibrium x_{ji}^*, $i = 1, 2, \ldots, I, j = 1, 2, \ldots, J$, is stable (see Saaty, 1981, p. 168).

2. *The Log-Linear Specification*

We now assume the following log-linear comparative advantages producing F functions

$$F_{ji} = A_{ji} \prod_{q=1}^{J} \prod_{p=1}^{I} x_{qp}(t)^{\alpha_{qp}^{ji}}, \qquad \text{(IV.A.2.1)}$$

so that for each i, k stocks located at j one has

$$\frac{x_{ji}(t+1)}{x_{jk}(t+1)} = \frac{A_{ji}}{A_{jk}} \prod_{q=1}^{J} \prod_{p=1}^{I} x_{qp}(t)^{(\alpha_{qp}^{ji} - \alpha_{qp}^{jk})}. \qquad \text{(IV.A.2.2)}$$

Obviously, from (IV.A.2.1), we have

$$\ln F_{ji} = \ln A_{ji} + \sum_{q=1}^{J} \sum_{p=1}^{I} \alpha_{qp}^{ji} \ln x_{qp}(t), \qquad \text{(IV.A.2.3)}$$

and

$$\frac{\partial \ln F_{ji}}{\partial x_{qp}(t)} = \frac{\alpha_{qp}^{ji}}{x_{qp}(t)}. \qquad \text{(IV.A.2.4)}$$

Thus, the Jacobi block-matrix of the stock/location effects includes the blocks

$$J_{pi}(t+1, t) = \|s_{qp}^{ji}(t+1, t)\|$$
$$= \left\| \frac{x_{ji}(t+1)}{x_{qp}(t)} \sum_{s=1}^{I} x_{js}(t+1)(\alpha_{qp}^{ji} - \alpha_{qp}^{js}) \right\|, \qquad q, j = 1, 2, \ldots, J. \qquad \text{(IV.A.2.5)}$$

At equilibrium, $x^*_{ji}, j = 1, 2, \ldots, J, i = 1, 2, \ldots, I$, we have

$$F^*_{ji} = A_{ji} \prod_q \prod_p (x^*_{qp})^{\alpha^{ji}_{qp}}, \tag{IV.A.2.6}$$

$$\frac{x^*_{ji}}{x^*_{jk}} = \frac{A_{ji}}{A_{jk}} \prod_q \prod_p (x^*_{qp})^{(\alpha^{ji}_{qp} - \alpha^{jk}_{qp})}, \tag{IV.A.2.7}$$

$$J^*_{pi} = \|s^{*ji}_{qp}\| = \left\| \frac{x^*_{ji}}{x^*_{qp}} \sum_{s=1}^{I} (\alpha^{ji}_{qp} - \alpha^{js}_{qp}) x^*_{js} \right\|. \tag{IV.A.2.8}$$

Again, the equilibrium is stable if and only if all eigenvalues of the Jacobi block-matrix are of absolute value less than one.

B. The Two-Stock, Two-Location Model

1. *The General Case*

The general case of the two-stock, two-location, discrete relative dynamics model is of the form

$$x_1(t+1) = \frac{F_1[x_1(t), x_2(t), y_1(t), y_2(t)]}{F_1[x_1(t), x_2(t), y_1(t), y_2(t)] + F_2[x_1(t), x_2(t), y_1(t), y_2(t)]}, \tag{IV.B.1.1}$$

$$x_2(t+1) = 1 - x_1(t+1) = \frac{F_2}{F_1 + F_2},$$

$$y_1(t+1) = \frac{H_1[x_1(t), x_2(t), y_1(t), y_2(t)]}{H_1[x_1(t), x_2(t), y_1(t), y_2(t)] + H_2[x_1(t), x_2(t), y_1(t), y_2(t)]}, \tag{IV.B.1.2}$$

$$y_2(t+1) = 1 - y_1(t+1) = \frac{H_2}{H_1 + H_2},$$

$$F_1, F_2, H_1, H_2 > 0.$$

For notational simplicity, we will present the model in the form

$$\begin{aligned} x(t+1) &= 1/1 + F[x(t), y(t)], \\ y(t+1) &= 1/1 + H[x(t), y(t)], \end{aligned} \tag{IV.B.1.3}$$

where $x(t+1) = x_1(t+1)$, $y(t+1) = y_1(t+1)$, $F = F_2/F_1$, $H = H_2/H_1$, F, $H > 0$. As a result of the above, we have

$$\frac{1 - x(t+1)}{x(t+1)} = F[x(t), y(t)], \qquad \frac{1 - y(t+1)}{y(t+1)} = H[x(t), y(t)]. \tag{IV.B.1.4}$$

There are four slopes organizing the Jacobi matrix

$$J(t+1,t) = \begin{bmatrix} s_{xx}(t+1,t) & s_{xy}(t+1,t) \\ s_{yx}(t+1,t) & s_{yy}(t+1,t) \end{bmatrix}, \qquad \text{(IV.B.1.5)}$$

where the entries are given by

$$\begin{aligned} s_{xx}(t+1,t) &= -x^2(t+1)\frac{\partial F}{\partial x(t)}, \\ s_{xy}(t+1,t) &= -x^2(t+1)\frac{\partial F}{\partial y(t)}, \\ s_{yx}(t+1,t) &= -y^2(t+1)\frac{\partial H}{\partial x(t)}, \\ s_{yy}(t+1,t) &= -y^2(t+1)\frac{\partial H}{\partial y(t)}. \end{aligned} \qquad \text{(IV.B.1.6)}$$

The trace and determinant of the Jacobi matrix (IV.B.1.5) are

$$\operatorname{Tr} J(t+1,t) = -x^2(t+1)\frac{\partial F}{\partial x(t)} - y^2(t+1)\frac{\partial H}{\partial y(t)}, \qquad \text{(IV.B.1.7)}$$

$$\det J(t+1,t) = x^2(t+1)y^2(t+1)\begin{vmatrix} \dfrac{\partial F}{\partial x(t)} & \dfrac{\partial F}{\partial y(t)} \\ \dfrac{\partial H}{\partial x(t)} & \dfrac{\partial H}{\partial y(t)} \end{vmatrix}. \qquad \text{(IV.B.1.8)}$$

At equilibrium, (x^*, y^*), the following conditions hold

$$\frac{1-x^*}{x^*} = F^*; \qquad \frac{1-y^*}{y^*} = H^*; \qquad \text{(IV.B.1.9)}$$

$$J^* = \begin{bmatrix} -x^*(1-x^*)\dfrac{\partial \ln F^*}{\partial x^*} & -x^*(1-x^*)\dfrac{\partial \ln F^*}{\partial y^*} \\ -y^*(1-y^*)\dfrac{\partial \ln H^*}{\partial x^*} & -y^*(1-y^*)\dfrac{\partial \ln H^*}{\partial y^*} \end{bmatrix}; \qquad \text{(IV.B.1.10)}$$

$$\operatorname{Tr} J^* = -x^*(1-x^*)\frac{\partial \ln F^*}{\partial x^*} - y^*(1-y^*)\frac{\partial \ln H^*}{\partial y^*}, \qquad \text{(IV.B.1.11)}$$

$$\Delta = \det J^* = x^*y^*(1-x^*)(1-y^*)\begin{vmatrix} \dfrac{\partial \ln F^*}{\partial x^*} & \dfrac{\partial \ln F^*}{\partial y^*} \\ \dfrac{\partial \ln H^*}{\partial x^*} & \dfrac{\partial \ln H^*}{\partial y^*} \end{vmatrix}. \qquad \text{(IV.B.1.12)}$$

The characteristic equation of the J^* matrix has the form

$$\lambda^2 - \operatorname{Tr} J^* + \Delta = 0,$$

and the stability conditions, requiring that $|\lambda_{1,2}| < 1$, mean that

$$-1 \pm \operatorname{Tr} J^* < \Delta < 1. \tag{IV.B.1.13}$$

2. *The Log-Linear Specifications*

Under log-linear specifications of the F functions, the two-stock, two-location model takes the form

$$\begin{aligned} x(t+1) &= 1/1 + F \\ &= 1/1 + Ax(t)^{a_{11}}[1 - x(t)]^{a_{12}} y(t)^{b_{11}}[1 - y(t)]^{b_{12}}, \\ y(t+1) &= 1/1 + H \\ &= 1/1 + Bx(t)^{a_{21}}[1 - x(t)]^{a_{22}} y(t)^{b_{21}}[1 - y(t)]^{b_{22}}, \end{aligned} \tag{IV.B.2.1}$$

where A and B are the environmental fluctuation (bifurcation) parameters, and the matrices of elasticities (the structural parameters) are

$$\begin{bmatrix} a_{11} & a_{12} \\ a_{21} & a_{22} \end{bmatrix}, \quad \begin{bmatrix} b_{11} & b_{12} \\ b_{21} & b_{22} \end{bmatrix}.$$

At equilibrium

$$\begin{aligned} \frac{\partial \ln F^*}{\partial x^*} &= \frac{a_{11} - (a_{11} + a_{12})x^*}{x^*(1 - x^*)}, \\ \frac{\partial \ln F^*}{\partial y^*} &= \frac{b_{11} - (b_{11} + b_{12})y^*}{y^*(1 - y^*)}, \\ \frac{\partial \ln H^*}{\partial x^*} &= \frac{a_{21} - (a_{21} + a_{22})x^*}{x^*(1 - x^*)}, \\ \frac{\partial \ln H^*}{\partial y^*} &= \frac{b_{21} - (b_{21} + b_{22})y^*}{y^*(1 - y^*)}. \end{aligned} \tag{IV.B.2.2}$$

Thus, from (IV.B.1.11, 12), the trace and determinant of the Jacobi matrix J are

$$\operatorname{Tr} J^* = -(a_{11} + b_{21}) + (a_{11} + a_{12})x^* + (b_{21} + b_{22})y^*, \tag{IV.B.2.3}$$

$$\begin{aligned} \Delta = \det J^* = &\begin{vmatrix} a_{11} & b_{11} \\ a_{21} & b_{21} \end{vmatrix} - x^* \begin{vmatrix} a_{11} + a_{12} & b_{11} \\ a_{21} + a_{22} & b_{21} \end{vmatrix} \\ &- y^* \begin{vmatrix} a_{11} & b_{11} + b_{12} \\ a_{21} & b_{21} + b_{22} \end{vmatrix} \\ &+ x^* y^* \begin{vmatrix} a_{11} + a_{12} & b_{11} + b_{12} \\ a_{21} + a_{22} & b_{21} + b_{22} \end{vmatrix}. \end{aligned} \tag{IV.B.2.4}$$

These explicit forms for the trace and determinant of the J^* matrix allow us to construct geometrically the domain of stability of the dynamic equilibrium

(x^*, y^*). In this case, the condition for stability (IV.B.1.13) obtains the form

$$-1 \pm [-(a_{11} + b_{21}) + (a_{11} + a_{12})x^* + (b_{21} + b_{22})y^*] < \Delta < 1. \tag{IV.B.2.5}$$

Geometrically, this inequality represents an area in the state variable space (x^*, y^*) bounded by three curves. Each of these three curves has an equation of the form

$$ax^*y^* - bx^* - cy^* = d, \tag{IV.B.2.6}$$

where a is a determinant

$$a = \begin{vmatrix} a_{11} + a_{12} & b_{11} + b_{12} \\ a_{21} + a_{22} & b_{21} + b_{22} \end{vmatrix}. \tag{IV.B.2.7}$$

If $a \neq 0$, then (IV.B.2.5) can be rewritten in the form of a rectangular hyperbola

$$\left(x^* - \frac{c}{a}\right)\left(y^* - \frac{b}{a}\right) = \frac{ad + bc}{a^2} \tag{IV.B.2.8}$$

referred to its asymptotes as axes, and with a center given by the point $(c/a, b/a)$. This implies that the domain of stability of the equilibrium point (x^*, y^*) is the intersection of the square, $0 \leq x^* \leq 1, 0 \leq y^* \leq 1$, and the areas between (or outside) the branches of three rectangular hyperbolas.

If, on the other hand, $a = 0$, then the boundaries of the domain of stability will be straight lines and the domain is a polygon, part of the unit square, $0 \leq x^*, y^* \leq 1$.

It is noted that the domain of stability for the equilibrium (x^*, y^*) does not depend upon the environmental fluctuation parameters (A, B).

3. *An Example*

Consider the model as specified by (IV.B.2.1), with positive A, B parameters, and the following elasticity (structural parameter) matrices

$$\begin{bmatrix} a_{11} & a_{12} \\ a_{21} & a_{22} \end{bmatrix} = \begin{bmatrix} 1 & 1 \\ -1 & 0 \end{bmatrix}; \quad \begin{bmatrix} b_{11} & b_{12} \\ b_{21} & b_{22} \end{bmatrix} = \begin{bmatrix} 1 & -1 \\ 1 & 0 \end{bmatrix}.$$

Then, according to (IV.B.2.3, 4) we obtain

$$\begin{gathered} \operatorname{Tr} J^* = -2 + 2x^* + y^*, \\ \Delta = \det J^* = 2x^*y^* - 3x^* - y^* + 2. \end{gathered} \tag{IV.B.3.1}$$

The conditions for stability are

$$-1 \pm (-2 + 2x^* + y^*) < 2x^*y^* - 3x^* - y^* + 2 < 1, \qquad 0 \leq x^*, \quad y^* < 1, \tag{IV.B.3.2}$$

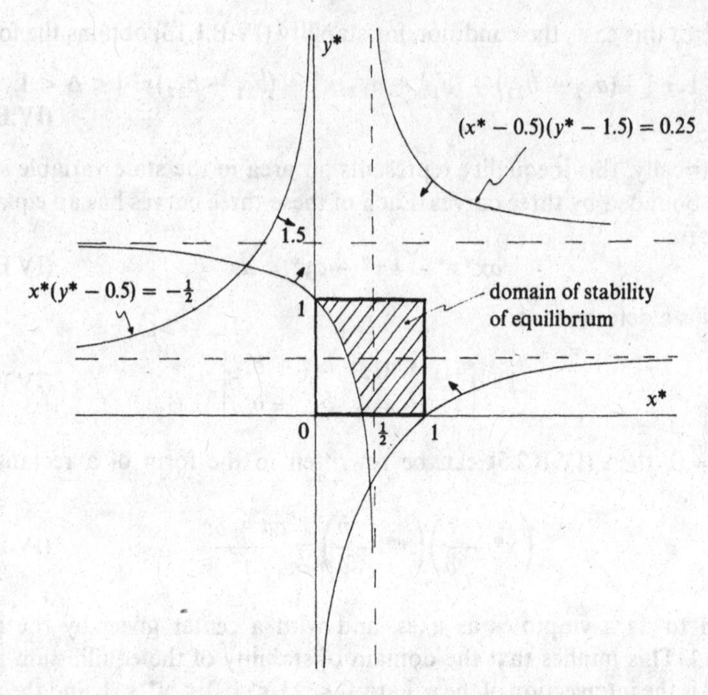

FIGURE 45. Domain of stability of the equilibrium for the special case of a log-linear, two-stock, two-location model. For parameters specifications, see text.

and will result in (see Figure 45)

$$x^*(y^* - 0.5) > -0.5, \qquad (x^* - 0.5)(y^* - 1.5) < 0.25, \qquad 0 \leq x^*, \quad y^* < 1. \tag{IV.B.3.3}$$

4. *Numerical Results for the Two-Stock, Two-Location, Log-Linear Model*

In Figure 46, a fixed-point behavior in both x and y is shown (a); a fixed point in y and a stable two-period cycle in x (b); and a stable two-period cycle in both x and y (c) are demonstrated. Parameter values responsible for these results are as follows: a fixed point in both x and y $(A = 10^{-2}, B = 1; a_{11} = -2, a_{12} = -1.5, b_{11} = -2, b_{12} = 0, a_{21} = 0.5, a_{22} = -1.5, b_{21} = 0, b_{22} = -1.5)$; one fixed point and a stable two-period cycle $(A = B = 10^{-2}, a_{11} = 0, a_{12} = -1.5, b_{11} = 1, b_{12} = 0, a_{21} = -2, a_{22} = -1, b_{21} = 1.5, b_{22} = 0)$; and the two stable two-period cycles $(A = B = 10^{-2}, a_{11} = -1, a_{12} = -1.5, b_{11} = 0.5, b_{12} = 0.5, a_{21} = -0.5, a_{22} = 0, b_{21} = -2, b_{22} = -2.)$

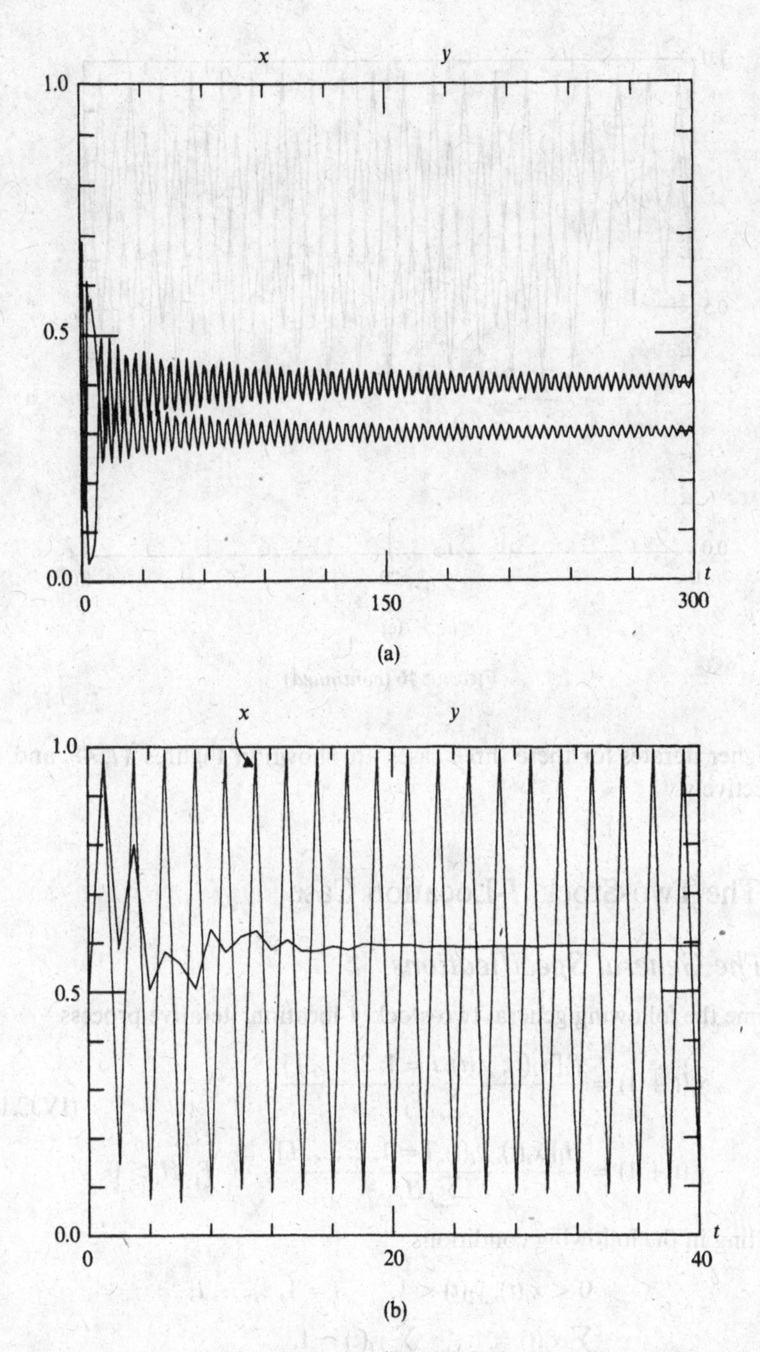

FIGURE 46. Various qualitative properties of equilibria for the two-stock, two-location, log-linear relative dynamics. A stable fixed point in x and y, a fixed point in y, a stable two-period cycle in x, and a stable two-period cycle in x and y. For specifications, see text.

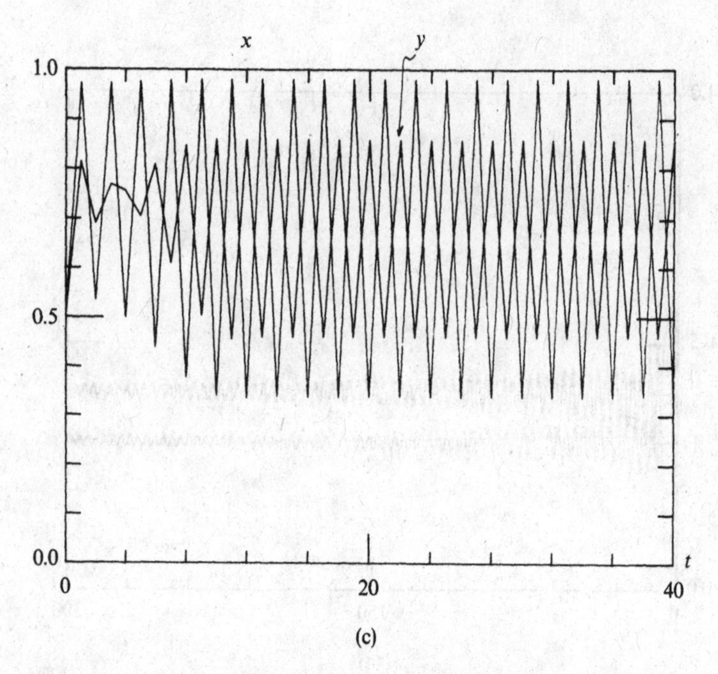

(c)

FIGURE 46 (*continued*)

Higher iterates for these three cases are shown in Figures 47, 48, and 49, respectively.

C. The Two-Stock, I-Location Case

1. *The General Specifications*

Assume the following general two-stock, I-location, iterative process

$$x_i(t+1) = \frac{F_i[x_i(t); y_i(t); i = 1, 2, \ldots, I]}{\sum_j F_j}, \tag{IV.C.1.1}$$

$$y_i(t+1) = \frac{H_i[x_i(t); y_i(t); i = 1, 2, \ldots, I]}{\sum_j H_j}, \qquad F_i, H_i > 0,$$

resulting in the following conditions

$$0 < x_i(t), y_i(t) < 1, \qquad i = 1, 2, \ldots, I,$$

$$\sum_i x_i(t) = 1; \quad \sum_i y_i(t) = 1.$$

FIGURE 47. Higher iterates of x and y in the two-stock, two-location model: two fixed points. First (a), second (b), and third (c) iterates of x, and first (d), second (e), and third (f) iterates of y plotted against x and y. ▷

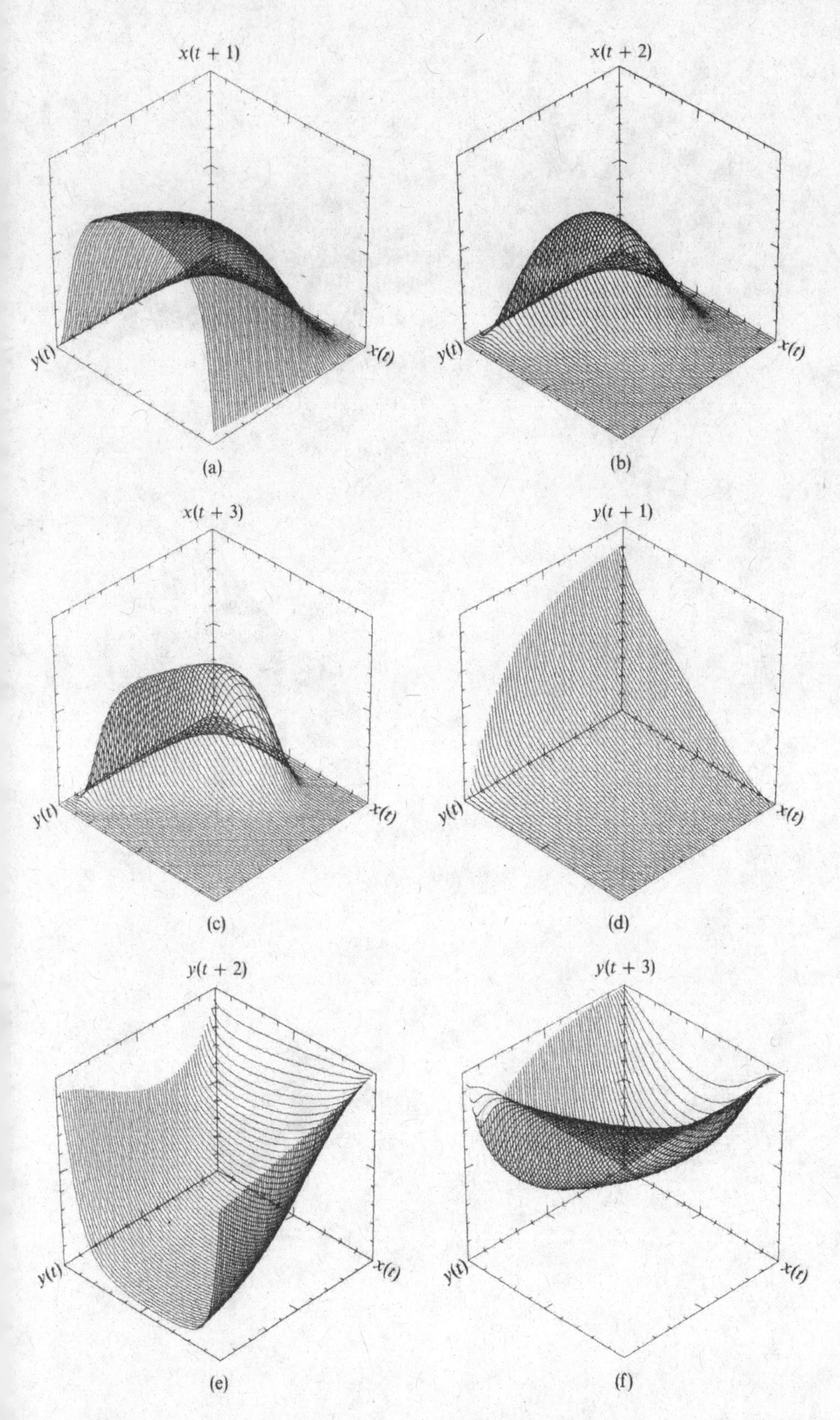
x(t + 1)
y(t)
x(t)
(a)
x(t + 2)
y(t)
x(t)
(b)
x(t + 3)
y(t)
x(t)
(c)
y(t + 1)
y(t)
x(t)
(d)
y(t + 2)
y(t)
x(t)
(e)
y(t + 3)
y(t)
x(t)
(f)

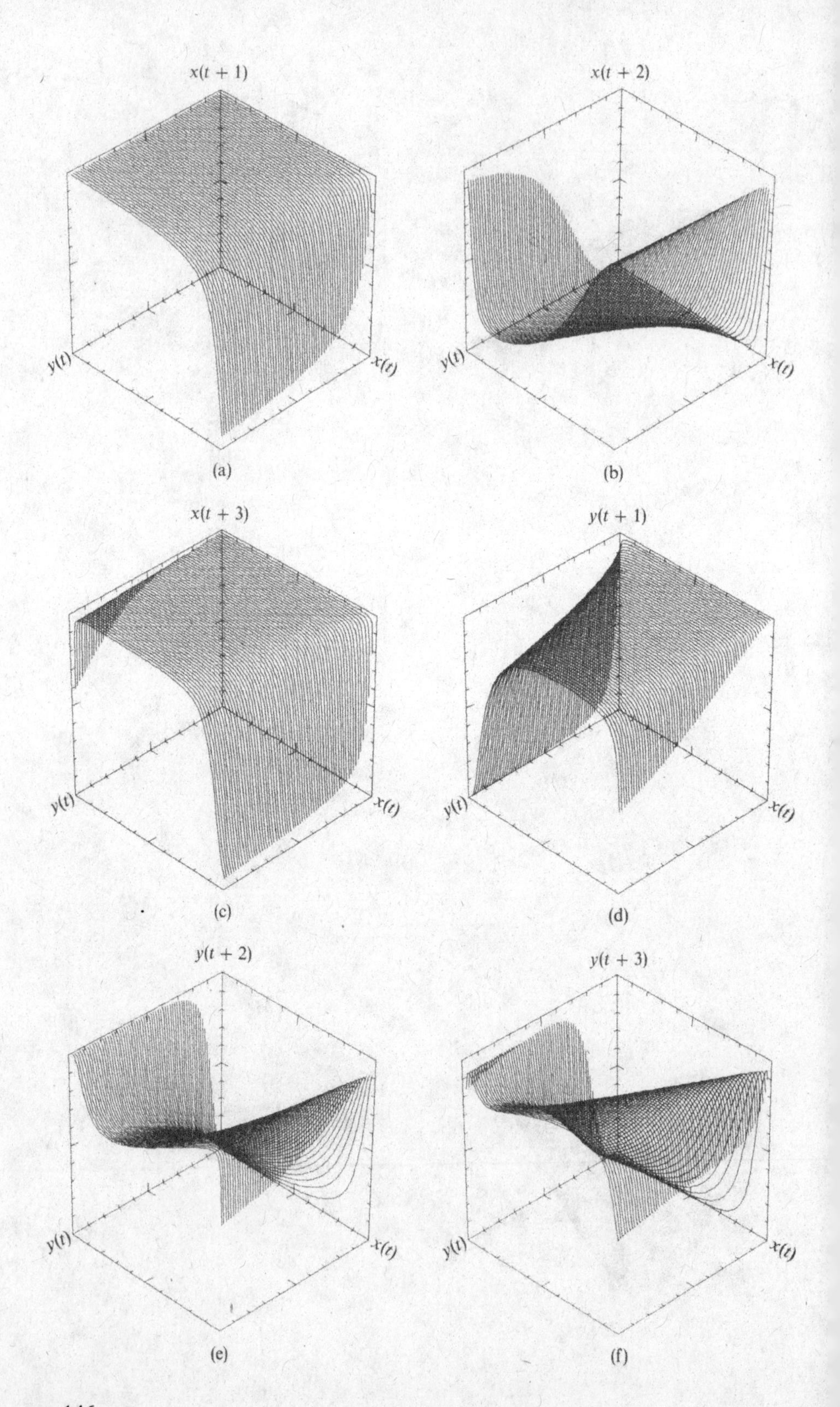
x(t + 1)
y(t)
x(t)
(a)
x(t + 2)
y(t)
x(t)
(b)
x(t + 3)
y(t)
x(t)
(c)
y(t + 1)
y(t)
x(t)
(d)
y(t + 2)
y(t)
x(t)
(e)
y(t + 3)
y(t)
x(t)
(f)

According to (IV.A.1.5, 6) the Jacobi block-matrix $J(t+1, t)$ has the following form

$$J(t+1,t) = \begin{bmatrix} J_{xx} & J_{xy} \\ J_{yx} & J_{yy} \end{bmatrix}, \tag{IV.C.1.2}$$

where

$$J_{xx} = \|s_{x_i x_j}(t+1,t)\| = \left\| \frac{\partial x_i(t+1)}{\partial x_j(t)} \right\|$$

$$= \left\| x_i(t+1) \sum_{\substack{s=1 \\ s \neq i}}^{I} x_s(t+1) \frac{\partial}{\partial x_j(t)} \ln \frac{F_i}{F_s} \right\|,$$

$$J_{xy} = \left\| x_i(t+1) \sum_{\substack{s=1 \\ s \neq i}}^{I} x_s(t+1) \frac{\partial}{\partial y_i(t)} \ln \frac{F_i}{F_s} \right\|, \tag{IV.C.1.3}$$

$$J_{yx} = \left\| y_i(t+1) \sum_{\substack{s=1 \\ s \neq i}}^{I} y_s(t+1) \frac{\partial}{\partial x_j(t)} \ln \frac{H_i}{H_s} \right\|,$$

$$J_{yy} = \left\| y_i(t+1) \sum_{\substack{s=1 \\ s \neq i}}^{I} y_s(t+1) \frac{\partial}{\partial y_j(t)} \ln \frac{H_i}{H_s} \right\|, \qquad i, j = 1, 2, \ldots, I.$$

At equilibrium $(x_1^*, x_2^*, \ldots, x_I^*; y_1^*, y_2^*, \ldots, y_I^*)$

$$x_i^* = \frac{F_i^*}{\sum_j F_j^*}, \qquad y_i^* = \frac{H_i^*}{\sum_j H_j^*}, \tag{IV.C.1.4}$$

or

$$x_i^* = 1/1 + \sum_{j \neq i} \frac{F_j^*}{F_i^*}; \qquad y_i^* = 1/1 + \sum_{j \neq i} \frac{H_j^*}{H_i^*}, \tag{IV.C.1.5}$$

or

$$\frac{1 - x_i^*}{x_i^*} = \frac{1}{F_i^*} \sum_{j \neq i} F_j^*, \qquad \frac{1 - y_i^*}{y_1^*} = \frac{1}{H_i^*} \sum_{j \neq i} H_j^*. \tag{IV.C.1.6}$$

There, the Jacobi block-matrix J^* includes the components

$$s_{x_i x_j}^* = x_i^* \sum_{s \neq i} x_s^* \frac{\partial}{\partial x_j^*} \ln \frac{F_i^*}{F_s^*},$$

$$s_{x_i y_i}^* = x_i^* \sum_{s \neq i} x_s^* \frac{\partial}{\partial y_i^*} \ln \frac{F_i^*}{F_s^*},$$

$$s_{y_i x_j}^* = y_i^* \sum_{s \neq i} y_s^* \frac{\partial}{\partial x_j^*} \ln \frac{H_i^*}{H_s^*}, \tag{IV.C.1.7}$$

$$s_{y_i y_j}^* = y_i^* \sum_{s \neq i} y_s^* \frac{\partial}{\partial y_i^*} \ln \frac{H_i^*}{H_s^*}, \qquad i, j = 1, 2, \ldots, I.$$

◁ FIGURE 48. Higher iterates of x and y in the two-stock, two-location model: fixed point in y, and a stable two-period cycle in x. First (a), second (b), and third (c) iterates of x, and first (d), second (e), and third (f) iterates of y plotted against x and y.

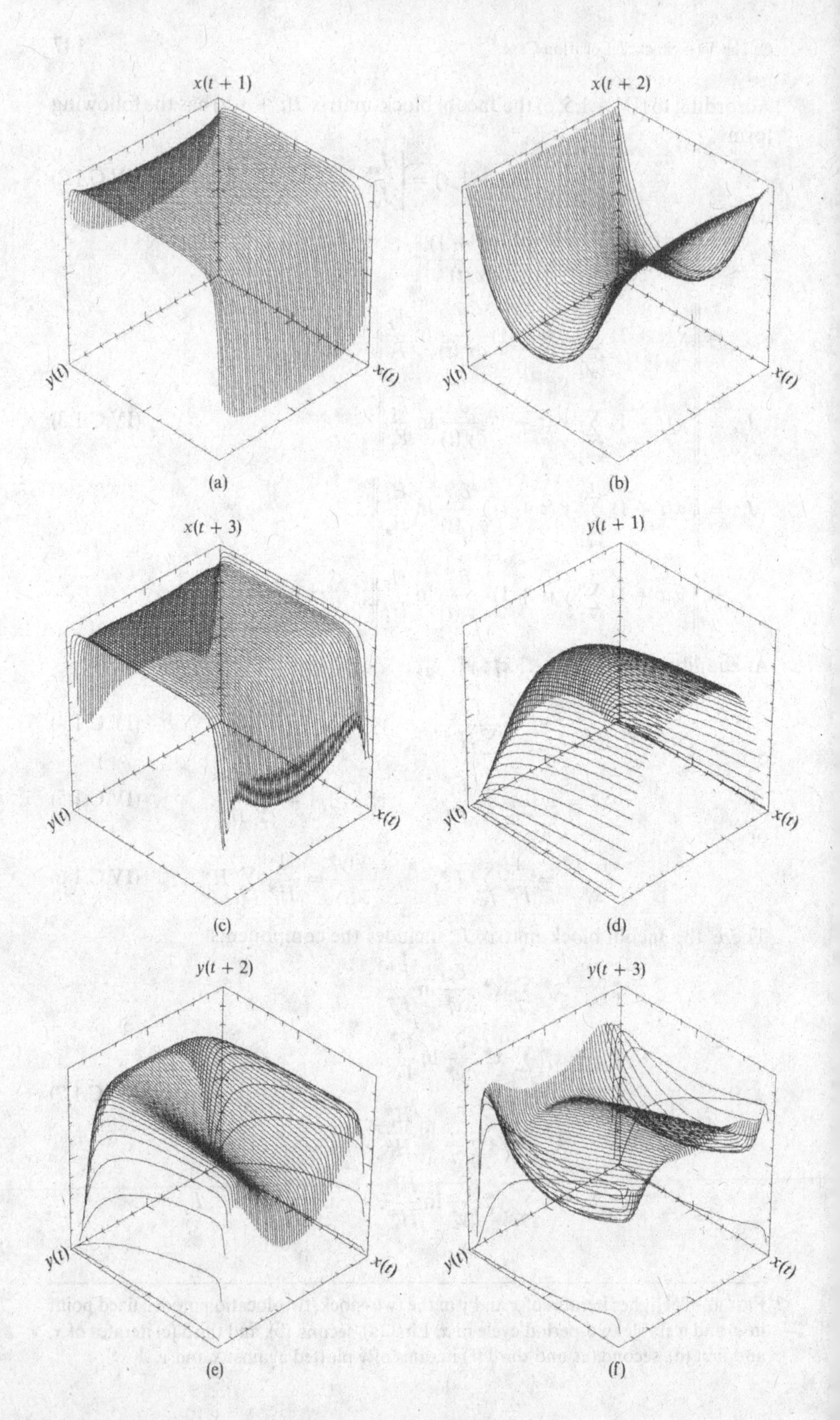
x(t + 1)
y(t)
x(t)
(a)
x(t + 2)
y(t)
x(t)
(b)
x(t + 3)
y(t)
x(t)
(c)
y(t + 1)
y(t)
x(t)
(d)
y(t + 2)
y(t)
x(t)
(e)
y(t + 3)
y(t)
x(t)
(f)

The inequality $|\lambda| < 1$, for each eigenvalue of the J^* matrix, gives the conditions for stability of the equilibrium $(x_1^*, x_2^*, \ldots, x_I^*; y_1^*, y_2^*, \ldots, y_I^*)$.

2. *The Log-Linear Specifications*

We now specify the F_i and H_i functions so that

$$F_i = A_i\left[\prod_{s=1}^{I} x_s(t)^{\alpha_{is}}\right]\left[\prod_{s=1}^{I} y_s(t)^{\beta_{is}}\right] = A_i \prod_{s=1}^{I} x_s(t)^{\alpha_{is}} y_x(t)^{\beta_{is}},$$

$$H_i = B_i\left[\prod_{s=1}^{I} x_s(t)^{\gamma_{is}}\right]\left[\prod_{s=1}^{I} y_s(t)^{\delta_{is}}\right] = B_i \prod_{s=1}^{I} x_s(t)^{\gamma_{is}} y_s(t)^{\delta_{is}}, \tag{IV.C.2.1}$$

$$A_i, B_i > 0; \qquad i = 1, 2, \ldots, I,$$

$$-\infty \le \alpha_{is}, \beta_{is}, \gamma_{is}, \delta_{is} \le +\infty, \qquad i, s = 1, 2, \ldots, I.$$

Assume a numéraire location I, then

$$\frac{x_i(t+1)}{x_I(t+1)} = \frac{F_i}{F_I} = \frac{A_i}{A_I}\prod_{s=1}^{I} x_s(t)^{(\alpha_{is}-\alpha_{Is})} y_s(t)^{(\beta_{is}-\beta_{Is})}; \qquad i = 1, 2, \ldots, I-1,$$

$$\frac{y_i(t+1)}{y_I(t+1)} = \frac{H_i}{H_I} = \frac{B_i}{B_I}\prod_{s=1}^{I} x_s(t)^{(\gamma_{is}-\gamma_{Is})} y_s(t)^{(\delta_{is}-\delta_{Is})}; \qquad i = 1, 2, \ldots, I-1. \tag{IV.C.2.2}$$

At equilibrium

$$x_i^* = \frac{F_i^*}{\sum_{j=1}^{I} F_j^*}; \qquad y_i^* = \frac{H_i^*}{\sum_{j=1}^{I} H_j^*}, \tag{IV.C.2.3}$$

$$x_i^* = x_I^* \frac{A_i}{A_I}\prod_{s=1}^{I} x_s^{*(\alpha_{is}-\alpha_{Is})} y_s^{*(\beta_{is}-\beta_{Is})},$$

$$y_i^* = y_I^* \frac{B_i}{B_I}\prod_{s=1}^{I} x_s^{*(\gamma_{is}-\gamma_{Is})} y_s^{*(\delta_{is}-\delta_{Is})}. \tag{IV.C.2.4}$$

The above can also be written as

$$\frac{A_I}{A_i} = x_i^{*(\alpha_{ii}-\alpha_{Ii}-1)} x_I^{*(\alpha_{iI}-\alpha_{II}+1)} \sum_{\substack{s=1 \\ s\neq i}}^{I-1} x_s^{*(\alpha_{is}-\alpha_{Is})} \prod_{s=1}^{I} y_s^{*(\beta_{is}-\beta_{Is})},$$

$$\frac{B_I}{B_i} = y_i^{*(\delta_{ii}-\delta_{Ii}-1)} y_I^{*(\delta_{iI}-\delta_{II}+1)} \prod_{s=1}^{I} x_s^{*(\gamma_{is}-\gamma_{Is})} \prod_{\substack{s=1 \\ s\neq i}}^{I-1} y_s^{*(\delta_{is}-\delta_{Is})}, \tag{IV.C.2.5}$$

◁ FIGURE 49. Higher iterates of x and y in the two-stock, two-location model: two stable two-period cycles. First (a), second (b), and third (c) iterates of x, and first (d), second (e), and third (f) iterates of y plotted against x and y.

and consequently, for $i = 1, 2, \ldots, I - 1$,

$$\ln \frac{A_I}{A_i} = (\alpha_{ii} - \alpha_{Ii} - 1) \ln x_i^* + (\alpha_{iI} - \alpha_{II} + 1) \ln x_I^* + \sum_{\substack{s=1 \\ s \neq i}}^{I-1} (\alpha_{is} - \alpha_{Is}) \ln x_s^* + \sum_{s=1}^{I} (\beta_{is} - \beta_{Is}) \ln y_s^*, \quad \text{(IV.C.2.6)}$$

$$\ln \frac{B_I}{B_i} = (\delta_{ii} - \delta_{Ii} - 1) \ln y_i^* + (\delta_{iI} - \delta_{II} + 1) \ln y_I^* + \sum_{s=1}^{I} (\gamma_{is} - \gamma_{Is}) \ln x_i^* + \sum_{\substack{s=1 \\ s \neq i}}^{I-1} (\delta_{is} - \delta_{Is}) \ln y_s^*. \quad \text{(IV.C.2.7)}$$

Thus, the system (IV.C.2.6, 7) is a linear one with $2(I - 1)$ equations and $2I$ unknown variables $\ln x_i^*, \ln y_i^*, i = 1, 2, \ldots, I$. In general, this sytem does not possess a solution that would satisfy the conservation conditions $\sum_i x_i^* = \sum_i y_i^* = 1$. One can, however, rewrite the system (IV.C.2.4) so that

$$\frac{x_i^*}{x_I^*} = \frac{A_i}{A_I} \left[\prod_{s=1}^{I-1} \left(\frac{x_s^*}{x_I^*} \right)^{(\alpha_{is} - \alpha_{Is})} \right] x_I^{* \sum_{s=1}^{I} (\alpha_{is} - \alpha_{Is})} \times \left[\prod_{s=1}^{I-1} \left(\frac{y_s^*}{y_I^*} \right)^{(\beta_{is} - \beta_{Is})} \right] y_I^{* \sum_{s=1}^{I} (\beta_{is} - \beta_{Is})}, \quad \text{(IV.C.2.8)}$$

$$\frac{y_i^*}{y_I^*} = \frac{B_i}{B_I} \left[\prod_{s=1}^{I-1} \left(\frac{x_s^*}{x_I^*} \right)^{(\gamma_{is} - \gamma_{Is})} \right] x_I^{* \sum_{s=1}^{I} (\gamma_{is} - \gamma_{Is})} \times \left[\prod_{s=1}^{I-1} \left(\frac{y_s^*}{y_I^*} \right)^{(\delta_{is} - \delta_{Is})} \right] y_I^{* \sum_{s=1}^{I} (\delta_{is} - \delta_{Is})}. \quad \text{(IV.C.2.9)}$$

If

$$\sum_{s=1}^{I} (\alpha_{is} - \alpha_{Is}) = 0, \qquad \sum_{s=1}^{I} (\gamma_{is} - \gamma_{Is}) = 0,$$
$$\sum_{s=1}^{I} (\beta_{is} - \beta_{Is}) = 0, \qquad \sum_{s=1}^{I} (\delta_{is} - \delta_{Is}) = 0, \quad \text{(IV.C.2.10)}$$

then, after converting to logarithms, one obtains the linear system of $2(I - 1)$ equations with $2(I - 1)$ unknowns, $\ln(x_i^*/x_I^*)$, $\ln(y_i^*/y_I^*)$, $i = 1, 2, \ldots, I - 1$. If the determinant of this system is not zero then a unique solution does exist. Moreover, if this solution is designated as

$$\ln\left(\frac{x_i^*}{x_I^*}\right) = a_i, \qquad \ln\left(\frac{y_i^*}{y_I^*}\right) = b_i, \qquad i = 1, 2, \ldots, I - 1, \quad \text{(IV.C.2.11)}$$

then the conservation conditions imply that

$$x_i^* = e^{a_i}/1 + \sum_{j=1}^{I-1} e^{a_j}; \qquad y_i^* = e^{b_i}/1 + \sum_{j=1}^{I-1} e^{b_j}, \qquad i = 1, 2, \ldots, I - 1,$$

$$x_I^* = 1/1 + \sum_{j=1}^{I-1} e^{a_j}; \qquad y_I^* = 1/1 + \sum_{j=1}^{I-1} e^{b_j}.$$

For the two-stock, multiple-location specifications of an universal map the following qualitative phenomena occur: unstable competitive exclusion; stable coexistence; 2^k-period cycles; local and global turbulence. However, many more interesting features of the universe of parameter space are still left largely to be explored, for example that of quasi-periodicity.

D. The Two-Stock, Multiple-Location, Discrete-Time, Logistic Growth Model

1. *Definitions and Analytical Properties*

In this section, we present briefly the two-species version of the model we outlined in Part III, Section E, namely the discrete-time equivalent of the continuous-time logistic growth model. A special interpretation of this model has been supplied by Sonis (1989) for labor-capital, spatiotemporal dependencies.

Consider the following specifications

$$
\begin{aligned}
x_i(t+1) &= \frac{y_i(t)F_i(t)}{\sum_{j=1}^{I} y_j(t)F_j(t)}, \\
y_i(t+1) &= \frac{x_i(t)H_i(t)}{\sum_{j=1}^{I} x_j(t)H_j(t)}, \\
0 \le x_i(t), y_i(t) &< 1, \qquad i = 1, 2, \ldots, I, \\
\sum_{j=1}^{I} x_i(t) &= \sum_{j=1}^{I} y_i(t) = 1, \\
0 < F_i(t) &= F_i[x_1(t), \ldots, x_I(t); y_1(t), \ldots, y_I(t)], \\
0 < H_i(t) &= H_i[x_1(t), \ldots, x_I(t); y_1(t), \ldots, y_I(t)].
\end{aligned}
\qquad \text{(IV.D.1.1)}
$$

A major result from this specification is that it does not allow for complete concentration of one stock, say population (x_i) at location i, and complete concentration of the other stock, say capital (y_j) at another location j. Only in the same location can the two stocks experience complete concentration. This will be shown next. It draws from Sonis (1989).

After appropriate substitutions from the original formulations, one obtains

$$
\begin{aligned}
x_i(t+1) &= \frac{x_i(t-1)F_i(t)H_i(t-1)}{\sum_{j=1}^{I} x_j(t-1)F_j(t)H_j(t-1)}, \\
y_i(t+1) &= \frac{y_i(t-1)F_i(t-1)H_i(t)}{\sum_{j=1}^{I} y_j(t-1)F_j(t-1)H_j(t)}.
\end{aligned}
\qquad \text{(IV.D.1.2)}
$$

This is a system of second-order difference equations with a two-time-period delay identifying interdependencies between these two stocks in space–time.

It is possible to present this system in discrete-time logistic growth form

$$x_i(t+1) = x_i(t-1) \sum_{s=1}^{I} x_s(t+1) \frac{F_i(t)H_i(t-1)}{F_s(t)H_s(t-1)},$$
$$y_i(t+1) = y_i(t-1) \sum_{s=1}^{I} y_s(t+1) \frac{F_i(t-1)H_i(t)}{F_s(t-1)H_s(t)}. \qquad \text{(IV.D.1.3)}$$

Further, for any pair of locations,

$$\frac{x_i(t+1)}{x_s(t+1)} = \frac{x_i(t-1)}{x_s(t-1)} \frac{F_i(t)}{F_s(t)} \frac{H_i(t-1)}{H_s(t-1)},$$
$$\frac{y_i(t+1)}{y_s(t+1)} = \frac{y_i(t-1)}{y_s(t-1)} \frac{F_i(t-1)}{F_s(t-1)} \frac{H_i(t)}{F_s(t)}, \qquad \text{(IV.D.1.4)}$$

and thus, for $t = 2\tau - 1$ ($\tau \geq 1$), one has

$$\frac{x_i(2\tau)}{x_s(2\tau)} = \frac{x_i(0)}{x_s(0)} \prod_{k=0}^{\tau-1} \frac{F_i[2(\tau-k)-1]}{F_s[2(\tau-k)-1]} \frac{H_i[2(\tau-k)-1]}{H_s[2(\tau-k)-1]}.$$

This implies that for $\tau \geq 1$, and under the conservation condition $\sum_i x_i(2\tau) = 1$,

$$x_i(2\tau) = \frac{x_i(0) \prod_{k=0}^{\tau-1} F_i[2(\tau-k)-1] H_i[2(\tau-k)-1]}{\sum_{j=1}^{I} x_j(0) \prod_{k=0}^{\tau-1} F_j[2(\tau-k)-1] H_j[2(\tau-k)-1]}. \qquad \text{(IV.D.1.5)}$$

Analogously

$$y_i(2\tau) = \frac{y_i(0) \prod_{k=0}^{\tau-1} F_i[2(\tau-k)-1] H_i[2(\tau-k)-1]}{\sum_{j=1}^{I} y_j(0) \prod_{k=0}^{\tau-1} F_j[2(\tau-k)-1] H_j[2(\tau-k)-1]}. \qquad \text{(IV.D.1.6)}$$

In the same manner, for $t = 2\tau$ ($\tau \geq 1$), one has with the help of conditions

$$x_i(1) = \frac{y_i(0)F_i(0)}{\sum_{j=1}^{I} y_j(0)F_j(0)},$$
$$y_i(1) = \frac{x_i(0)H_i(0)}{\sum_{j=1}^{I} x_j(0)H_j(0)},$$

the complete description of logistic growth in discrete dynamics

$$x_i(2\tau+1) = \frac{y_i(0)F_i(0) \prod_{k=0}^{\tau-1} F_i[2(\tau-k)] H_i[2(\tau-k)-1]}{\sum_{j=1}^{I} y_j(0)F_j(0) \prod_{k=0}^{\tau-1} F_j[2(\tau-k)] H_j[2(\tau-k)-1]},$$
$$y_i(2\tau+1) = \frac{x_i(0)H_i(0) \prod_{k=0}^{\tau-1} F_i[2(\tau-k)-1] H_i[2(\tau-k)]}{\sum_{j=1}^{I} x_j(0)H_j(0) \prod_{k=0}^{\tau-1} F_j[2(\tau-k)-1] H_j[2(\tau-k)]}. \qquad \text{(IV.D.1.7)}$$

If the comparative advantage functions, F_i, H_i, are independent of the relative size of the two stocks as distributed to the I locations, then the generalized (standard), logistic two-species, multinomial logistic, growth model is obtained

(Sonis, 1989)

$$F_i = A_i, \qquad H_i = B_i, \qquad i = 1, 2, \ldots, I,$$

$$x_i(2\tau) = \frac{x_i(0)(A_iB_i)^\tau}{\sum_{j=1}^I x_j(0)(A_jB_j)^\tau}, \tag{IV.D.1.8}$$

$$y_i(2\tau) = \frac{y_i(0)(A_iB_i)^\tau}{\sum_{j=1}^I y_j(0)(A_iB_i)^\tau},$$

and

$$x_i(2\tau+1) = \frac{y_i(0)A_i^{\tau+1}B_i^\tau}{\sum_{j=1}^I y_j(0)A_j^{\tau+1}B_j^\tau},$$

$$y_i(2\tau+1) = \frac{x_i(0)A_i^\tau B_i^{\tau+1}}{\sum_{j=1}^I x_j(0)A_j^\tau B_j^{\tau+1}}. \tag{IV.D.1.9}$$

This interdependence between the two stocks, *each stock's initial perturbation* alone affecting the *intra*locational dynamics of the other stock, is of interest. It implies that the change in relative abundance in one stock at some location is proportional to the original value of the other stock at that location.

2. *The Jacobi Block-Matrix*

The above specifications provide the basis for fully describing the two-stock relative dynamics near the competitive exclusion equilibrium state. According to conditions (IV.C.1.2) the four blocks of the Jacobi block-matrix $J(t+1, t)$ have the form

$$J_{xx} = \|s_{x_ix_j}\|, \quad J_{xy} = \|s_{x_iy_i}\|, \quad J_{yx} = \|s_{y_ix_j}\|, \quad J_{yy} = \|s_{y_iy_j}\|. \tag{IV.D.2.1}$$

Here each component is of the form

$$s_{x_ix_j} = x_i(t+1)\left[\sum_{\substack{s=1\\s\neq i}}^I x_s(t+1)\frac{\partial}{\partial x_j(t)}\ln\frac{F_i(t)}{F_s(t)}\right], \qquad i \neq j,$$

$$s_{x_iy_i} = x_i(t+1)\left[\sum_{\substack{s=1\\s\neq i}}^I x_s(t+1)\frac{\partial}{\partial y_i(t)}\ln\frac{F_i(t)}{F_s(t)}\right] + [1 - x_i(t+1)]\frac{F_i(t)}{\sum_s y_s(t)F_s(t)},$$

$$s_{x_iy_j} = x_i(t+1)\left[\sum_{\substack{s=1\\s\neq i}}^I x_s(t+1)\frac{\partial}{\partial y_j(t)}\ln\frac{F_i(t)}{F_s(t)}\right] - \frac{x_i(t+1)F_j(t)}{\sum_s y_s(t)F_s(t)}, \qquad i \neq j,$$

$$s_{y_ix_i} = y_i(t+1)\left[\sum_{\substack{s=1\\s\neq i}}^I y_s(t+1)\frac{\partial}{\partial x_i(t)}\ln\frac{H_i(t)}{H_s(t)}\right] + [1 - y_i(t+1)]\frac{H_i(t)}{\sum_s x_s(t)H_s(t)},$$

$$s_{y_ix_j} = y_i(t+1)\left[\sum_{\substack{s=1\\s\neq i}}^I y_s(t+1)\frac{\partial}{\partial x_j(t)}\ln\frac{H_i(t)}{H_s(t)}\right] - \frac{y_i(t+1)H_j(t)}{\sum_s x_s(t)H_s(t)},$$

$$s_{y_iy_j} = y_i(t+1)\left[\sum_{\substack{s=1\\s\neq i}}^I y_s(t+1)\frac{\partial}{\partial y_j(t)}\ln\frac{H_i(t)}{H_s(t)}\right]. \tag{IV.D.2.2}$$

3. *Dynamic Structure of the Equilibria*

At equilibrium $(x_1^*, x_2^*, \ldots, x_I^*; y_1^*, y_2^*, \ldots, y_I^*)$ the following conditions are met

$$x_i^* = \frac{y_i^* F_i^*}{\sum_{j=1}^I y_j^* F_j^*}; \qquad y_i^* = \frac{x_i^* H_i^*}{\sum_{j=1}^I x_j^* H_j^*}. \tag{IV.D.3.1}$$

The important feature of this model is the simultaneity in competitive exclusion, i.e., the simultaneous presence (concentration) or absence of both stocks at any location i. This is directly a result of the symmetry found in the conditions $x_i^* = 0$ and $y_i^* = 0$ for any i in I. It is easily checked that simultaneous complete or partial concentrations at some location(s) r of both stocks is a stable dynamic equilibrium (stable global or local competitive exclusion).

To prove this, we consider the competitive exclusion equilibria

$$x_r^* = y_r^* = 1; \qquad x_i^* = y_i^* = 0, \qquad i \neq r, \tag{IV.D.3.2}$$

and check the components of the Jacobi block-matrix J^*

$$\begin{aligned}
&s_{x_i x_j}^* = 0, && s_{y_i y_j}^* = 0, && i, j = 1, 2, \ldots, I,\\
&s_{x_i y_r}^* = 0, && s_{y_i x_r}^* = 0, && i = 1, 2, \ldots, I,\\
&s_{x_r y_j}^* = -\frac{F_j^*}{F_r^*}; && s_{y_r x_j}^* = -\frac{H_j^*}{H_r^*}; && j \neq r,\\
&s_{x_i y_i}^* = \frac{F_i^*}{F_r^*}; && s_{y_i x_i}^* = \frac{H_i^*}{H_r^*}; && i \neq r.
\end{aligned} \tag{IV.D.3.3}$$

Therefore, the characteristic polynomial for the eigenvalues of the Jacobi matrix J^* has the form $\det(\lambda I - J^*) = 0$ where the matrix $\lambda I - J^*$ has four square $(I \times I)$ submatrices

$$\lambda I - J^* = \left[\begin{array}{c|c} Q_1 & F \\ \hline H & Q_2 \end{array}\right], \tag{IV.D.3.4}$$

where

$$Q_1 = Q_2 = \begin{bmatrix} \lambda & 0 & \ldots & 0 \\ 0 & \lambda & \ldots & 0 \\ \vdots & \vdots & & \vdots \\ 0 & 0 & \ldots & \lambda \end{bmatrix},$$

and

$$F = \begin{bmatrix} -\frac{F_1^*}{F_r^*} & 0 & \ldots & 0 \\ 0 & -\frac{F_2^*}{F_r^*} & \ldots & 0 \\ \vdots & \vdots & & \vdots \\ \frac{F_1^*}{F_r^*} & \frac{F_2^*}{F_r^*} & \ldots & \frac{F_I^*}{F_r^*} \\ \vdots & \vdots & & \vdots \\ 0 & 0 & \ldots & -\frac{F_I^*}{F_r^*} \end{bmatrix} \begin{matrix} \\ \\ \\ r\text{th row}, \\ \\ \\ \end{matrix}$$

and equivalently for H with the substitution $H \to F$. Matrix J^* at the competitive exclusion condition has the following eigenvalues

$$0;\ 0;\ = \pm\sqrt{\frac{F_j^* H_j^*}{F_r^* H_r^*}}, \qquad j \neq r, \quad j = 1, 2, \ldots, I.$$

Conditions for stability require that $|\lambda| < 1$, implying the following

$$F_r^* H_r^* > F_j^* H_j^*, \qquad j \neq r, \quad j = 1, 2, \ldots, I, \tag{IV.D.3.5}$$

for complete (global) competitive exclusion. For partial (local) competitive exclusion equilibria of the form

$$\begin{aligned} x_{r_s}^* \neq 0, \qquad y_{r_s}^* \neq 0, \qquad s = 1, 2, \ldots, p; \\ x_i^* = y_i^* = 0, \qquad i \neq r_s, \end{aligned} \tag{IV.D.3.6}$$

consider the reduced Jacobi $p \times p$ matrix

$$J_{r_1, r_2, \ldots, r_p}^* = \left[\begin{array}{ccc|ccc} S_{x_{r_1} x_{r_1}}^* & \cdots & S_{x_{r_1} x_{r_p}}^* & S_{x_{r_1} y_{r_1}}^* & \cdots & S_{x_{r_1} y_{r_p}}^* \\ \vdots & & \vdots & \vdots & & \vdots \\ S_{x_{r_p} x_{r_1}}^* & \cdots & S_{x_{r_p} x_{r_p}}^* & S_{x_{r_p} y_{r_1}}^* & \cdots & S_{x_{r_p} y_{r_p}}^* \\ \hline S_{y_{r_1} x_{r_1}}^* & \cdots & S_{y_{r_1} x_{r_p}}^* & S_{y_{r_1} y_{r_1}}^* & \cdots & S_{y_{r_1} y_{r_p}}^* \\ \vdots & & \vdots & \vdots & & \vdots \\ S_{y_{r_p} x_{r_1}}^* & \cdots & S_{y_{r_p} x_{r_p}}^* & S_{y_{r_p} y_{r_1}}^* & \cdots & S_{y_{r_p} y_{r_p}}^* \end{array}\right].$$

Conditions for stability of the equilibria (IV.D.3.6) include two types of requirements:

(i) that the absolute value of each eigenvalue of the reduced Jacobi matrix $J_{r_1, r_2, \ldots, r_p}^*$ be less than 1; and

(ii)
$$\begin{aligned} \max_i (F_i^* H_i^*) &= F_{r_1}^* H_{r_1}^* = F_{r_2}^* H_{r_2}^* = \cdots \\ &= F_{r_p}^* H_{r_p}^* > F_j^* H_j^*, \qquad j \neq r_1, \quad r_2, \ldots, r_p. \end{aligned} \tag{IV.D.3.7}$$

Conclusions

With this part, we conclude the analytical and numerical treatment of our universal map of spatiotemporal, discrete-time, discrete-space, stock interdependencies and growth in relative abundance. Numerical results seem to indicate that the region of dynamic stability shrinks rapidly around the core of the parameter space, where all structural parameters of the composite advantage generating functions are zero.

Stable and/or unstable, global or local, competitive exclusion seems to be a dominant feature of these model specifications. They are exclusively, almost, present in the case of a discrete-time version of the continuous-time logistic growth model. Clearly, much more extensive numerical (and analytical) treatment of this algorithm is needed to fully grasp the variety of performance it contains at these dimensions.

Epilogue

In spite of the many interesting features of the map uncovered, either by analysis or computer simulation, many items need further elaboration. For example, the mathematical proofs for most of our findings were not presented, many still awaiting the mathematician's attention. Computer simulations were kept at a relatively coarse level. The full application potential of the universal map in socio-spatial dynamic theory was not explored. Empirical evidence supplied was indeed limited.

The purpose of this book is not to carry out in full any of these tasks. Its main objective is to demonstrate the various dynamical features of the universal map of relative discrete-time dynamics. In this final section of the book we will first take stock of the map; we will then suggest ways of using it in a number of socioeconomic–geographical areas of research. Finally, we will supply a few suggestions for further research.

1. *Six Central Issues*

In assessing the elements of potential utility such a map possesses, six items will be highlighted: the spatiotemporal slaving property; simple and calm dynamics; the variety of cycles it is possible to model; complexity and turbulent behavior; various types of chae; and the aspect of "renormalization."

a. Spatio Temporal Slaving

Physicists (e.g., Haken, 1981) have coined the phrase "slaving principle" to classify slow and fast changes, and their underlying determinism: slow moving variables, according to this principle, "slave" the behavior of fast changing variables. When one is interested in the dynamics of fast moving (state) variables, then one can consider the slow moving ones as constants, i.e., as (fixed) parameters.

Evolutionary, geological, and cosmological time are used in biology, geology, and astrophysics, respectively, and they identify different time constants. We introduce the notion of "sociological" time to identify changes in social stocks recorded in human generations' time. The latter still involve fast and slow changes, although these still might be for observers of social stocks relatively very fast when compared with the previous three types of time constants.

Interdependencies among stocks and locations are not only dynamic (manifested in time) and subject to the temporal slaving principle; they are also manifested in space and are subject to spatial slaving. The spatial slaving principle is present when certain variables (simply because they operate in macrospace) influence, without being influenced by other variables operating in microspace. Similarly to the temporal slaving principle, spatial slaving implies that certain (microspatially invariant) variables operate at a macrospatial scale, i.e., they vary only when one moves over relatively large chunks of space. Whereas, there are other variables which are spatially varying, i.e., they change when one moves over small areas of space. Consideration of the slaving (macrospace) variables can be suppressed when the focus is on the microspace scale, i.e., on the spatially slaved and varying variables.

Whether linear or nonlinear interdependencies are included in modeling social systems is very much a function of the spatial and temporal horizon under consideration. Linear interdependencies are only approximations acceptable over relatively short spatiotemporal frames. Long-term spatial or temporal analysis must be carried out in a nonlinear mode. The decision to adopt a linear or nonlinear model is thus tied to the spatiotemporal horizon and the degree of complexity required by the model.

There are nine possible types of models that can be constructed according to the classifications shown in Table 7. At one end of the spectrum, model type A, one includes only the fast moving slaved variables of a small scale space. At the other end of the spectrum, model type I, spatially and temporally highly complex models are constructed where changes in both slaving and slaved variables are included. Intermediate complexity models are covered by the cases B–H.

The dichotomy between fast and slow moving variables need not be sharp, as intermediate rates of change in both state variables and parameters may be present. In our map, the environmental parameter A was assumed to vary faster than the structural parameters (i.e., the exponents of the F functions). Conversely, qualitative properties of stability may dictate the model's structural complexity and its appropriate space–time framework.

TABLE 7. The presence of the slaving principle in space–time.

		Temporal term		
		Short	Long	Short & long
Spatial extent	Micro	*A*	B	C
	Macro	*D*	*E*	*F*
	Micro and Macro	*G*	*H*	*I*

b. Simple and Calm Dynamics

A dominant type of simple and calm dynamics, involving smooth and continuous change in socioeconomic–geographic systems, is logistic growth and/or decline. Logistic-type dynamics, of the variety shown to be possible within the realm of the universal map, might be rather easy to document in social histories–events. Logistic growth and/or decline are indeed ubiquitous events in the demographic, economic, and geographic literature. Exponential population increase or decrease has been widely documented in socio-spatial dynamics, the most celebrated being the Malthusian population growth model.

One can safely contend that simple and calm dynamics are the norm for socio-spatial systems, frequently observed in many locations and time periods. Mostly they occur over short distances (spatial neighborhoods) and short time horizons (temporal vicinities).

Linear models are special versions of nonlinear ones. Models with unique stable (spiral or nodal) dynamic equilibrium are also special formulations of more complex nonlinear forms. Numerous socio-spatial dynamic events, recorded on a variety of socioeconomic variables and documented with data from recent history and at many locations, are adequately depicted by such models. The universal map contains all these types of dynamics, as demonstrated in Parts I through IV.

c. Cycles

As already mentioned, linear interdependencies are only approximations acceptable over relatively short spatiotemporal frames. Broadening the scope of inquiry requires that long-term (spatial or temporal) analysis be carried out in a nonlinear mode. The decision to adopt a linear or nonlinear model is thus partly tied to the degree of spatiotemporal complexity required by the model. Demand for a model to replicate, regularly and irregularly, periodic and nonperiodic cycles may require a special type of nonlinear model capable of reproducing all these dynamics.

In studying cycles one must distinguish between irregular oscillations in the state variables, which could be the outcome of random changes in the model's parameters, and regular or irregular oscillations, which are caused by inherent fast dynamics associated with specific parameter values. This two-fold origin of cycles not only presents severe statistical difficulties, but it also taxes the analyst's speculative capacity in studying them.

To address the above issues a theory of socio-spatial cycles must be hypothesized. This is not the appropriate forum to do so, except to supply a few elements that such a theory must contain. Associated with different environmental changes or fluctuations, social stocks and their events exhibit different behavior, in space–time. Changes in a location's topographic, geologic, or climatic features may periodically occur. They may obey very long or very short time-spans relative to sociological time. They may be characterized by highly regular motion (seasonable variability being a case in point), or they may reveal irregular, sharp, and sudden changes (earthquakes are an example of such catastrophic events). These movements may cause social variables to oscillate with frequencies ranging from a few months to millennia, and from a few square miles to global events.

Economic variables like wealth, income, prices, profits, etc., and business cycles associated with capital accumulation or decumulation, are examples of relatively short-term and irregularly periodic cycles occurring with frequencies ranging from one to ten years. There are social cycles of a much higher frequency and/or regularity. For example, daily cycles are recorded in the rhythm by which roads or buildings are used in 24-hour intervals. Hourly, daily, or even shorter time-period cycles are recorded in human physiology; yearly cycles are found in nomadic and migratory patterns of human populations in many regions of the earth.

At the other end of the spectrum, one can find century-long irregular cycles, or even highly infrequent cycles with a period extending over millennia. Very large scale human migration movements are examples of such cycles. Social, economic, demographic, cultural, and technological forces determine the frequencies of these regular or irregular cycles. One expects demographic forces to determine longer frequency cycles than economic factors, with macro-cultural and major technological innovation forces generating socio-spatial cycles of even longer frequencies, extending over much larger space.

One might detect cycles in political events, or other social dynamics. For example, the size of groups following various political, religious, moral, or ideological views may obey oscillatory behavior in space–time of human history–geography.

The universal map proposed provides the framework within which such motions of different time–space scales can be accommodated, and a general theory for socio-spatial cycles can be formulated on the basis of locational–temporal advantages involving time delays. Many speeds can be accounted for by the map. In the dynamic behavior of the state variables the fastest movements can be recorded. Various speeds of change can be accounted for in the movement of the bifurcation (environmental) parameters. Even slower modification of the environment can be depicted by the model's structural parameters.

Evolutionary processes may be captured by appropriate modifications to the universal map's entire form, whereby the model becomes more or less complex, with a higher or lower dimensionality. This topic will not be explored, here, however.

d. Complex and Turbulent Dynamics

Among the many calm and simple dynamics in social systems behavior, there are moments and locations of violent, complex, and possibly chaotic movement. We suggest that it is these unusual events which cause evolution of social stocks.

Violent episodes of the peculiar type found in the turbulent behavior of the universal map may be temporally and spatially rare social events: and they may be few and far between. Chaos may be a relatively inaccessible state in both space and time. An observer of social systems must be very fortunate to stumble onto social science data where the presence of turbulence is captured.

Chaotic events may not only be rare but even unique, i.e., nonrecursive events which are rather brief in the history and geography of social systems. Uniqueness and/or briefness make temporal and/or spatial proximity to turbulent regimes very difficult, when ticking occurs in real time (or space). As critical points in the parameter space are approached, i.e., thresholds where period-doubling cycles split or the various chaotic motions start or bifurcate, the number of iterations (ticks or generations) it takes for a stock to reach the equilibrium state approaches asymptotically infinity. If the recording of social events requires a period or location of stability this factor may inhibit turbulent events from being recorded.

Further, social systems may often be kept away from violent events (spatially and/or temporally) because of the built-in delay mechanism in reaching those states in real time and/or space. These delay mechanisms may be attributed to various social, political, demographic, economic, or other forces responsible for the iterative (delay) form of the model. This built-in inaccessibility could be overridden by appropriate environmental fluctuations pushing the system from calm to violent regimes. The circumstance, however, must be very particular for this to occur. Fluctuations must be of specific magnitude and push toward a particular direction in each instance for turbulent events to occur, both conditions making it rather unlikely.

Contrary to the calmness of social stock dynamics is the behavior of stock prices in stock exchanges. Prices being special cases of relative stocks are susceptible to the dynamic events portrayed in this model. Analyses of stock price behavior, or indices of bundles of stocks in reference to a futures market index or other appropriate indicators, are likely to reveal past turbulent patterns.

Physical social stocks with certain temporal and spatial durability and fixity, like human population, built capital stock, etc., may lack such properties over relatively short time periods. Very high transporation and/or transaction costs are behind such fixity. Ephemerally behaving stocks, as diverse as stock prices, interest rates, money supply and the like in economics, and human emotions and perceptions in psychology, may not be very calm on the other hand.

e. CHAE

Much of the earlier work on turbulence in the 1970s and early 1980s focused on specific forms of turbulence and chaotic motion, drawn largely from the earlier work by May (1976), Feigenbaum (1978), (1979), (1980), and Lorenz (1963). This map is a departure from the earlier work, for a number of reasons, some of which were presented in Parts I through IV. A key reason, however, is that the universal map demonstrated the possibility that different types of turbulence exist. Thus, a new perspective on chaos emerges for socio-spatial and other systems.

Regular periodic movement and the slope-sequences found in period-doubling cycles of the universal map may be events that future social scientists may have to await millennia of recorded social histories to observe, if ever at all. What is more likely to record (or have already been recorded although we might not be aware of it) is nonperiodic (chaotic) motion in much shorter time-spans. Even more likely, what we might have been observing is a sequence of transitions among different chaotic motions. This realization may be of some use to historians, and in the fields of archeology and cultural anthropology. It could also be of interest to psychologists studying chaotic changes in the levels of many variables at the individual level, and to mass psychologists interested in modeling turbulence in mass emotions and perceptions at an aggregate (collective) level.

f. Renormalization

In the abstraction of the one-stock, multiple-location, universal map we identified the state variables as "human population" distributed in discrete space. We suggest that the map proposed contains the possible dynamics of a broad array of social stocks. Each of them, however, may occupy only a specific area in the abstract parameter space, and in the dimensionality of the map. Whatever the specific interpretation of the variables or parameters one wishes to assign to them, the map's dynamics are inclusive.

Different regions of the map's parameter space in the n-dimensional state variable problem are associated with qualitatively different dynamics. Based on these properties one may assign a particular problem to a particular region of the parameter space for a specific model dimension n^*.

The model's universality is also attributed to the maximum number of qualitatively different dynamic events it contains, capable of replicating any dynamic behavior of any social stock. Within this "structuralist" framework one views the previous work by the authors, e.g., Dendrinos (1980), (1984a), Sonis (1986a).

One can extend this conjecture and state that natural and social systems may occupy different positions in the abstract parameter space of the universal map, for different dimensionality n. Clearly, dynamics of human population size (and other variables such as prices, love, or fear) may occupy different regions in the map's dimensionality and associated parameter space than in particular molecular dynamics. This may be the substantive meaning of the renormalization process found in physics (K.G. Wilson, 1971).

2. *Three Areas of Application*

We now proceed to pointing out particular areas of socioeconomic–geographic research in which the universal map can extend existing static or linear dynamic models. Three specific extensions are shown: the input–output model; the gravitational interaction model; and the discrete choice decision-making theory.

A number of innovative dynamic features, introduced into the current analysis, are investigated to an extent. Further elaboration is left, however, to future detailed research.

a. Dynamics and Input–Output Analysis

Consider the input–output (I–O) flow square matrix $[x_{ij}]$, $i, j = 1, 2, \ldots, I$, where i designates stock type and j stands for process type. Each entry, x_{ij}, designates the quantity of input stock i flowing into production/consumption process j at some time t. Thus, the I–O flow matrix is expressed in terms of input flows. At this stage there is no need to consider multiple output types per production–consumption process. Neither is there need to differentiate among primary and intermediate stocks, or among production–consumption processes (intermediate and final).

Entries of the I–O matrix are expressed in quantities, so that divisions of the row totals

$$X_i = \sum_j x_{ij}$$

are possible and produce the (input) coefficients

$$0 < a_{ij} = \frac{x_{ij}}{X_i} < 1,$$

$$\sum_{j=1}^{I} a_{ij} = 1; \qquad i = 1, 2, \ldots, I.$$

Under these specifications, summation over i of the entries x_{ij} for each j is meaningless, and so are similar summations of the (input) coefficients a_{ij}. Quantity X_i is the total amount of input i available at t. If Y_j is the total output of the production–consumption process j at t, then

$$0 < a_{ih} = \frac{x_{ih}}{X_i} = \frac{x_{ih}}{Y_j} < 1; \qquad j = i, \quad h = 1, 2, \ldots, I,$$

since at t

$$X_i = Y_j; \qquad i = j, \quad i, j = 1, 2, \ldots, I,$$

for all input types.

The universal map is fit to address any relative (and static) quantity distributed among the alternatives (like the coefficients a_{ij}) whose sum over these alternatives is equal to unity. Export base theory, locational quotient analysis, etc., are other prime examples of such cases.

In the above, merely accounting I–O (static) process, there is no excess quantity of input–output. Each production or consumption process produces–consumes one output j, simultaneously being the quantity of input i ($i = j$) employed at t. This is the I–O accounting conservation condition.

The I–O coefficients are derived by dividing the quantity of input i employed in process j by the total quantity of output in process j, i.e., a_{ij} are not only the input coefficients but also the I–O coefficients in the quantity conserving accounting I–O process.

When prices are considered, then the I–O model can be expressed in terms of value of input type i flowing into process j, so then

$$q_{ij} = p_i x_{ij},$$

where p_i is the price of stock i at t. Summations are now allowed vertically and horizontally

$$Q_{i.} = \sum_j q_{ij} = \sum_j p_i x_{ij} = p_i \sum_j x_{ij} = p_i X_i,$$

$$Q_{.j} = \sum_i q_{ij} = \sum_i p_i x_{ij},$$

$$Q_{i.} = Q_{.j}, \qquad i = j,$$

$$\sum_i Q_{i.} = \sum_i p_i x_i = \sum_j Q_{.j} = \bar{Q}.$$

Average cost in I–O models equals marginal cost, and both are equal to the price of the input factor i; profits are zero for all production–consumption processes.

From the above, one can derive the quantity $Q_{.j}$ for any known (or targeted) level E_i as a function of the prices and I–O coefficients

$$Q = P(I - A)^{-1} E,$$

where P is the vector of input prices, I is the identity matrix, A is the I–O coefficients matrix, and E is the fixed at t column of requirements to be met by the I–O process.

Leontief's extension of the static I–O accounting model is a linear dynamic model of the form

$$X = (I - A)^{-1} E - B\dot{X},$$

where $\dot{X}$ represents the time derivative of the vector of inputs and B is an $I \times I$ matrix of coefficients, exogenous to the production/consumption process. An alternative to Leontief's (1970) dynamic formulation is that by Hannon (1985), whereby the matrix B is a diagonal matrix still exogenous to the I–O process.

In both models, only fast dynamics are recorded, i.e., only the state variables X change following the system's dynamic specifications. Both matrices of coefficients, A and B, remain unchanged by assuming either (endogenous to the model) variable elasticities of substitution or technological progress. This is a significant drawback to conventional I–O analysis.

Using the universal map, an alternative formulation will be supplied here. A nonlinear dynamic I–O model will be formulated; fast dynamics in the state variables and slow changes in the production/consumption process, i.e., the I–O coefficients, will be directly linked. However, the benefits of introducing these dynamics has a cost, in that the model's dimensionality in the parameter space increases exponentially.

Assuming that

$$a_{ij}(t+1) = \frac{F_{ij}(t)}{\sum_{j=1}^{I} F_{ij}(t)},$$

$$F_{ij}(t) = C_{ij} \prod_h a_{ih}(t)^{\alpha_{ih}^j} > 0,$$

where C_{ij} are $I \times I$ nonnegative coefficients and α_{ih}^j are I^3 real numbers. The coefficients C_{ij} are the environmental relatively slow changing parameters; whereas the exponents α_{ih}^j are the very slowly adjusting comparative elasticities

$$\alpha_{ih}^j = \frac{\partial F_{ij}(t)}{\partial a_{ih}} \bigg/ F_{ij}(t),$$

in the production/consumption process. Variables $F_{ij}(t)$ identify temporal comparative advantages in the flow of stock i into process j in view of the alternatives for using the input stock. The benefit of introducing the endogenous dynamics, as defined by the universal map, is the extraordinary gamut of performance obtained; but these benefits are countered by the costs of having to deal with $I(I^2 - 1)$ parameters. Thus, nonlinear dynamics require a much larger set of parameters than the Leontief (I^2) and Hannon (I) parameter formulatons.

b. Gravitational Interaction, Information Theory, and the Map

Consider the matrix of static flows comprising a trip distribution tableau $[x_{ij}]$ such that

$$\sum_i \sum_j x_{ij}(t) = 1; \qquad i = 1, 2, \ldots, I, \quad j = 1, 2, \ldots, J,$$

where the entries of the $I \times J$ matrix represent relative size flows from origin i to destination j at time t, giving rise to a probability vector ($0 < x_{ij} < 1$). Further, consider that the only factor determining these flows is spatial impedance given by an impedance tableau $[c_{ij}]$.

i. *The Static Unconstrained Gravity Model of Probabilistic Flows*

According to conventional formulations found in the spatial literature, the form of the unconstrained static gravity model (without, that is, any origin or destination constraints) depends on a positive transportation cost ($c_{ij} > 0$) count so that

$$x_{ij}(t) = k(t) c_{ij}(t)^{-\beta},$$

$$k(t) = \left[\sum_i \sum_j c_{ij}(t)^{-\beta} \right]^{-1},$$

where $k(t)$ is a gravitational constant guaranteeing that the entries of the distribution tableau are probabilities. Parameter β is a sensitivity to travel impedance measure.

ii. *The Static Single-Constraint Gravity Model*

Under this specification two possible models can be constructed, an origin-constraint and a destination-constraint gravity model

$$x_{ij}(t) = k_i(t) X_i(t) c_{ij}(t)^{-\beta},$$

$$k_i(t) = \left[\sum_j c_{ij}(t)^{-\beta} \right]^{-1} \quad \text{so that} \quad \sum_i X_i(t) = 1;$$

or

$$x_{ij}(t) = k_j(t) X_j(t) c_{ij}(t)^{-\beta},$$

$$k_j(t) = \left[\sum_i c_{ij}(t)^{-\beta} \right]^{-1} \quad \text{so that} \quad \sum_j X_j(t) = 1.$$

The first formulation represents the origin-constraint gravity formulation, whereas the second statement stands for the destination-constraint static gravity model.

iii. *The Doubly Constrained Gravity Model*

With a set of both origin and destination constraints, the static gravity model now looks like

$$x_{ij}(t) = k_i(t) k_j(t) X_i(t) X_j(t) c_{ij}(t)^{-\beta},$$

$$k_i(t) = \left[\sum_j k_j(t) X_j(t) c_{ij}(t)^{-\beta} \right]^{-1},$$

$$k_j(t) = \left[\sum_i k_i(t) X_i(t) c_{ij}(t)^{-\beta} \right]^{-1},$$

so that

$$\sum_i X_i(t) = \sum_j X_j(t) = 1.$$

The information theory-based entropy measure of the above trip distribution is

$$H(t) = -\sum_i \sum_j x_{ij}(t) \ln x_{ij}(t),$$

or

$$H_1(t) = -\sum_i \sum_j k(t) c_{ij}(t)^{-\beta} \ln k(t) c_{ij}(t)^{-\beta},$$

under formulation (i), or

$$H_{2.1}(t) = -\sum_i \sum_j [k_i(t) X_i(t) c_{ij}(t)^{-\beta}] \ln [k_i(t) X_i(t) c_{ij}(t)^{-\beta}],$$

$$H_{2.2}(t) = -\sum_i \sum_j [k_j(t) X_j(t) c_{ij}(t)^{-\beta}] \ln [k_j(t) X_j(t) c_{ij}(t)^{-\beta}],$$

under formulation (ii), or

$$H_{3(t)} = -\sum_i \sum_j [k_i(t)k_j(t)X_i(t)X_j(t)c_{ij}(t)^{-\beta}] \ln[k_i(t)k_j(t)X_i(t)X_j(t)c_{ij}(t)^{-\beta}]$$

under formulation (iii).

iv. *Iterative Gravitational Interaction*

Let us now introduce in the above gravitational iteration model an iterative process, by making the future expected flow a function of the currently perceived transport cost; and by making the spatial impedance, i.e., the transportation costs matrix, a function of relative spatial flows. These additions are done by incorporating a congestion cost function

$$c_{ij}(t) = c_{ij}^0\left[1 + \alpha\left(\frac{x_{ij}(t)}{z_{ij}}\right)^{\gamma}\right],$$

where $c_{ij}(t)$ is the currently experienced transportation costs from origin i to destination j, c_{ij}^0 is the uncongested cost on the path from i to j, z_{ij} is the relative capacity of the path, and (α, γ) are positive parameters. The iterative process can be stated as follows, under specification (i),

$$x_{ij}(t+1) = k(t)\left\{c_{ij}^0\left[1 + \alpha\left(\frac{x_{ij}(t)}{z_{ij}}\right)^{\gamma}\right]\right\}^{-\beta},$$

$$k(t) = \left[\sum_i \sum_j \left\{c_{ij}^0\left[1 + \alpha\left(\frac{x_{ij}(t)}{z_{ij}}\right)^{\gamma}\right]\right\}^{-\beta}\right]^{-1}.$$

Turbulent dynamics may be present under this lagged response in spatial flows, which may or may not have a stable (fixed-point) dynamic equilibrium. The new (at $t + 1$) entropy count is

$$H(t+1) = -\sum_i \sum_j [k(t+1)c_{ij}(t+1)^{-\beta}] \ln[k(t+1)c_{ij}(t+1)^{-\beta}]$$

and its dynamics may be chaotic!

c. Individual Discrete-Choice Models

Assume that at time t an individual i, belonging to a homogeneous group of I individuals, chooses (spatial) alternative j ($j = 1, 2, \ldots, J$) with a probability given by

$$p_{ij}(t) = \frac{U_{ij}(t)}{\sum_j U_{ij}(t)} = \frac{1}{1 + \sum_{k \neq j}\left[\frac{U_{ik}(t)}{U_{ij}(t)}\right]}.$$

In the above statement, the probability $p_{ij}(t)$ is a function of a set of relative advantages enjoyed by individual i from alternative j compared with all other alternatives. This comparative advantage is a utility ratio: $U_{ik}(t)/U_{ij}(t)$.

Further, assume that each utility level $U_{ij}(t)$ is a log-linear function of a bundle of attributes found in alternative j

$$U_{ij}(t) = A_{ij} \prod_m x_{jm}(t)^{\alpha^i_m},$$

so that

$$\frac{U_{ik}(t)}{U_{ij}(t)} = \frac{A_{ik}}{A_{ij}} \prod_m \left[\frac{x_{km}(t)}{x_{jm}(t)}\right]^{\alpha^i_m}.$$

One can make the additional assumption that, when spatial choices are involved where congestion externalities are present, the attribute level $x_{km}(t)$ is affected by the number of individuals opting for it. Thus, in a simple manner, we can assume

$$x_{km}(t) = a_{km} p_{ik}(t)^{\beta_m}.$$

The comparative advantages then of alternative j for an individual i when compared with all other alternatives are

$$p_{ij}(t) = 1/1 + \sum_{k \neq j} \frac{A_{ik}}{A_{ij}} \prod_m \frac{a_{km}}{a_{jm}} \left[\frac{p_{ik}(t)}{p_{ij}(t)}\right]^{\alpha^i_m \beta_m},$$

so that the attraction/repulsion of an alternative j is a function of the congested use of the various other options compared with j.

Iterative dynamics are so introduced, as the future $(t + 1)$ expected probability, and the then realized distribution $p_{ij}(t + 1)$ is based on the currently (i.e., at t) perceived congestion costs (comparative advantages) of spatial alternative j

$$p_{ij}(t + 1) = 1/1 + \sum_{k \neq j} A'_{ik} \prod_m a'_{km} \left[\frac{p_{ik}(t)}{p_{ij}(t)}\right]^{\gamma^i_m},$$

where

$$A'_{ik} = \frac{A_{ik}}{A_{ij}}; \qquad a'_{km} = \frac{a_{km}}{a_{jm}}; \qquad \gamma^i_m = \frac{\alpha^i_m}{\beta_m}.$$

Again, the discrete dynamics of spatial choices involving a time lag between expected spatial choice and currently perceived utility of alternatives may contain turbulence.

3. *Further Research Suggestions*

Many items were not elaborated on in full here. For example, the mathematical proofs for most of our findings, analytical and computational, were kept at a minimal level. The numerical simulation results were too coarse. Broad epistemological aspects of our abstract socio-spatial dynamic theory were not fully addressed. Empirical evidence supplied was limited. The areas suggested for applying the algorithm could be expounded on and extended. Potentially, all areas of social science modeling involving probabilities may provide grounds for incorporating the universal map.

One could argue, however, that subjects such as these can never be fully addressed in a research volume like this one. To the critical reader who may take issue with our choice we point to the innovative aspects of this book: the new mathematical results of the map, both analytical and numerical, and the framework for setting up an abstract and comprehensive theory of socio-spatial evolution based on temporal and locational advantages of alternatives involving time lags in choices.

In all, the innovative aspects of this book and the detail of their elaboration had to be traded off. The present work is not the complete statement on either count. Among the many immediate extensions, elaborating on the analytical results (assisted by a less coarse numerical analysis) is of high priority. More extensive computer simulations of the I-stock, J-location problem present the possibility of obtaining new and interesting findings. Poincaré sections, circle maps, and rotation number analysis could provide more insights into the dynamic behavior of the universal map.

On the empirical side, the appropriate recording of histories and the ex-post description of stock market price behavior also open ground for fruitful extensions. Finally, the presentation of a speculative (i.e., iterative, lags involving, dynamic) theory in a number of social sciences based on the universal map may hide a significant amount of new insights and understanding in all of them.

The faculties of the universal map can be expanded further than we have attempted to stress. An interesting perspective, suggested to us by Professor Kawashima of Gakushuin University, is to look at the map from the viewpoint of t representing discrete space and i (the state variable's subscript standing for type of alternative or stock) designating time. Or, to incorporate explicitly both discrete (lagged) time and space in the model

$$x_j(t+1, r) = \frac{A_j F_j[x(t, r)]}{\sum_h A_h F_h[x(t, r)]}, \qquad h, j = 1, 2, \ldots, J, \quad t = 1, 2, \ldots, T,$$

$$x_j(t, r+1) = \frac{B_j G_j[x(t, r)]}{\sum_h B_h G_h[x(t, r)]}, \qquad r = 1, 2, \ldots, R,$$

where the subscript j is an index of stock type; and T and R are the time and space horizons, respectively. One may analyze a number of variations of this map, e.g., the case where $x_j(t+1, r)$ and $x_j(t, r+1)$ depend on both F and G functions. Substantive interpretations of such formulations extending our model could also be of interest.

Along the F function specifications there are many other forms open to examination, besides the log-linear and exponential functions analyzed here. For instance, F could be specified as

$$F_i(t) = A_i \left[\sum_j x_j(t)^{\alpha_{ij}} \right]^{-\beta}, \qquad \beta > 0.$$

The form of the above functions closely resembles those of congested trip flows discussed in a previous section (2b) of the Epilogue.

In this volume the map's behavior was analyzed under the restriction that the initial stock values at all locations were less than one and greater than zero. One could also analyze the cases of these values falling in domains outside this relative size region, where $0 < x(t) < 1$. Under specific integer exponents cases could be analyzed where $x(0) < 0$. Doing so, the global behavior of the state variables can be uncovered and some new dynamic phenomena may be found.

The basis for deriving the map was the locational–temporal advantages enjoyed in the spatial alternatives open to a social stock, coupled with a time delay (discrete dynamics) reaction mechanism. It is suggested that this formulation trenscends specific disciplines within social sciences. An area where this universal map can be directly applied is demography. The dynamics of relative shares of population contained in various age cohorts can be effectively captured along the lines suggested here. Thus, various specifications of the map can capture the different ecological associations among various cohorts in space–time. It may also extend its applicability into the biological and natural sciences. If so, an effort toward addressing substantive differences among these disciplines or sciences, giving rise to methodological similarities under the framework of the universal map, may lay down the foundation for a unification of these fields of inquiry.

Appendix I
Second-Order Determinants of the Three-Location, One-Stock Model

The determinants Δ_1, Δ_2, Δ_3 of equation (III.A.3.4), giving the size of Δ in (III.A.3.2), are supplied by the expressions

$$\Delta_1 = \begin{vmatrix} s_{11}^* & s_{12}^* \\ s_{21}^* & s_{22}^* \end{vmatrix}$$

$$= \begin{vmatrix} \dfrac{x_1^*}{F_1}\left[\dfrac{\partial F_1^*}{\partial x_1^*} - x_1^*\left(\dfrac{\partial F_1^*}{\partial x_1^*} + \dfrac{\partial F_2^*}{\partial x_1^*} + \dfrac{\partial F_3^*}{\partial x_1^*}\right)\right], & \dfrac{x_1^*}{F_1^*}\left[\dfrac{\partial F_1^*}{\partial x_2^*} - x_1^*\left(\dfrac{\partial F_1^*}{\partial x_2^*} + \dfrac{\partial F_2^*}{\partial x_2^*} + \dfrac{\partial F_3^*}{\partial x_2^*}\right)\right] \\ \dfrac{x_2^*}{F_2^*}\left[\dfrac{\partial F_2^*}{\partial x_1^*} - x_2^*\left(\dfrac{\partial F_1^*}{\partial x_1^*} + \dfrac{\partial F_2^*}{\partial x_1^*} + \dfrac{\partial F_3^*}{\partial x_1^*}\right)\right], & \dfrac{x_2^*}{F_2^*}\left[\dfrac{\partial F_2^*}{\partial x_2^*} - x_2\left(\dfrac{\partial F_1^*}{\partial x_2^*} + \dfrac{\partial F_2^*}{\partial x_2^*} + \dfrac{\partial F_3^*}{\partial x_2^*}\right)\right] \end{vmatrix},$$

$$\Delta_2 = \begin{vmatrix} s_{11}^* & s_{13}^* \\ s_{31}^* & s_{33}^* \end{vmatrix}$$

$$= \begin{vmatrix} \dfrac{x_1^*}{F_1^*}\left[\dfrac{\partial F_1^*}{\partial x_1^*} - x_1^*\left(\dfrac{\partial F_1^*}{\partial x_1^*} + \dfrac{\partial F_2^*}{\partial x_1^*} + \dfrac{\partial F_3^*}{\partial x_1^*}\right)\right], & \dfrac{x_1^*}{F_1^*}\left[\dfrac{\partial F_1^*}{\partial x_3^*} - x_1^*\left(\dfrac{\partial F_1^*}{\partial x_3^*} + \dfrac{\partial F_2^*}{\partial x_3^*} + \dfrac{\partial F_3^*}{\partial x_3^*}\right)\right] \\ \dfrac{x_3^*}{F_3^*}\left[\dfrac{\partial F_3^*}{\partial x_1^*} - x_3^*\left(\dfrac{\partial F_1^*}{\partial x_1^*} + \dfrac{\partial F_2^*}{\partial x_1^*} + \dfrac{\partial F_3^*}{\partial x_1^*}\right)\right], & \dfrac{x_3^*}{F_3^*}\left[\dfrac{\partial F_3^*}{\partial x_3^*} - x_3^*\left(\dfrac{\partial F_1^*}{\partial x_3^*} + \dfrac{\partial F_2^*}{\partial x_3^*} + \dfrac{\partial F_3^*}{\partial x_3^*}\right)\right] \end{vmatrix},$$

$$\Delta_3 = \begin{vmatrix} s_{22}^* & s_{23}^* \\ s_{32}^* & s_{33}^* \end{vmatrix}$$

$$= \begin{vmatrix} \dfrac{x_2^*}{F_2^*}\left[\dfrac{\partial F_2^*}{\partial x_2^*} - x_2^*\left(\dfrac{\partial F_1^*}{\partial x_2^*} + \dfrac{\partial F_2^*}{\partial x_2^*} + \dfrac{\partial F_3^*}{\partial x_2^*}\right)\right], & \dfrac{x_2^*}{F_2^*}\left[\dfrac{\partial F_2^*}{\partial x_3^*} - x_2^*\left(\dfrac{\partial F_1^*}{\partial x_3^*} + \dfrac{\partial F_2^*}{\partial x_3^*} + \dfrac{\partial F_3^*}{\partial x_3^*}\right)\right] \\ \dfrac{x_3^*}{F_3^*}\left[\dfrac{\partial F_3^*}{\partial x_2^*} - x_3^*\left(\dfrac{\partial F_1^*}{\partial x_2^*} + \dfrac{\partial F_2^*}{\partial x_2^*} + \dfrac{\partial F_3^*}{\partial x_2^*}\right)\right], & \dfrac{x_3^*}{F_3^*}\left[\dfrac{\partial F_3^*}{\partial x_3^*} - x_3^*\left(\dfrac{\partial F_1^*}{\partial x_3^*} + \dfrac{\partial F_2^*}{\partial x_3^*} + \dfrac{\partial F_3^*}{\partial x_3^*}\right)\right] \end{vmatrix}.$$

Further,

$$\Delta_1 = \frac{x_1^*}{F_1^*} \cdot \frac{x_2^*}{F_2^*} \begin{vmatrix} \dfrac{\partial F_1^*}{\partial x_1^*} & \dfrac{\partial F_1^*}{\partial x_2^*} \\ \dfrac{\partial F_2^*}{\partial x_1^*} & \dfrac{\partial F_2^*}{\partial x_2^*} \end{vmatrix} - \left(\frac{\partial F_1^*}{\partial x_1^*} + \frac{\partial F_2^*}{\partial x_1^*} + \frac{\partial F_3^*}{\partial x_1^*}\right) \begin{vmatrix} x_1^* & \dfrac{\partial F_1^*}{\partial x_2^*} \\ x_2^* & \dfrac{\partial F_2^*}{\partial x_2^*} \end{vmatrix} - \left(\frac{\partial F_1^*}{\partial x_2^*} + \frac{\partial F_2^*}{\partial x_2^*} + \frac{\partial F_3^*}{\partial x_2^*}\right) \begin{vmatrix} \dfrac{\partial F_1^*}{\partial x_1^*} & x_1^* \\ \dfrac{\partial F_2^*}{\partial x_1^*} & x_2^* \end{vmatrix},$$

$$\Delta_2 = \frac{x_1^*}{F_1^*} \cdot \frac{x_3^*}{F_3^*} \begin{vmatrix} \frac{\partial F_1^*}{\partial x_1^*} & \frac{\partial F_1^*}{\partial x_3^*} \\ \frac{\partial F_3^*}{\partial x_1^*} & \frac{\partial F_3^*}{\partial x_3^*} \end{vmatrix} - \left(\frac{\partial F_1^*}{\partial x_1^*} + \frac{\partial F_2^*}{\partial x_1^*} + \frac{\partial F_3^*}{\partial x_1^*}\right) \begin{vmatrix} x_1^* & \frac{\partial F_1^*}{\partial x_3^*} \\ x_3^* & \frac{\partial F_3^*}{\partial x_3^*} \end{vmatrix} - \left(\frac{\partial F_1^*}{\partial x_3^*} + \frac{\partial F_2^*}{\partial x_3^*} + \frac{\partial F_3^*}{\partial x_3^*}\right) \begin{vmatrix} \frac{\partial F_1^*}{\partial x_1^*} & x_1^* \\ \frac{\partial F_3^*}{\partial x_1^*} & x_3^* \end{vmatrix},$$

$$\Delta_3 = \frac{x_2^*}{F_2^*} \cdot \frac{x_3^*}{F_3^*} \begin{vmatrix} \frac{\partial F_2^*}{\partial x_2^*} & \frac{\partial F_2^*}{\partial x_3^*} \\ \frac{\partial F_3^*}{\partial x_2^*} & \frac{\partial F_3^*}{\partial x_3^*} \end{vmatrix} - \left(\frac{\partial F_1^*}{\partial x_2^*} + \frac{\partial F_2^*}{\partial x_2^*} + \frac{\partial F_3^*}{\partial x_2^*}\right) \begin{vmatrix} x_2^* & \frac{\partial F_2^*}{\partial x_3^*} \\ x_3^* & \frac{\partial F_3^*}{\partial x_3^*} \end{vmatrix} - \left(\frac{\partial F_1^*}{\partial x_3^*} + \frac{\partial F_2^*}{\partial x_3^*} + \frac{\partial F_3^*}{\partial x_3^*}\right) \begin{vmatrix} \frac{\partial F_2^*}{\partial x_2^*} & x_2^* \\ \frac{\partial F_3^*}{\partial x_2^*} & x_3^* \end{vmatrix}.$$

Therefore,

$$\Delta = \left(\frac{x_1^*}{F_1^*}\right)^2 \left\{ \begin{vmatrix} \frac{\partial F_1^*}{\partial x_1^*} & \frac{\partial F_1^*}{\partial x_2^*} \\ \frac{\partial F_2^*}{\partial x_1^*} & \frac{\partial F_2^*}{\partial x_2^*} \end{vmatrix} + \begin{vmatrix} \frac{\partial F_1^*}{\partial x_1^*} & \frac{\partial F_1^*}{\partial x_3^*} \\ \frac{\partial F_3^*}{\partial x_1^*} & \frac{\partial F_3^*}{\partial x_3^*} \end{vmatrix} + \begin{vmatrix} \frac{\partial F_2^*}{\partial x_2^*} & \frac{\partial F_2^*}{\partial x_3^*} \\ \frac{\partial F_3^*}{\partial x_2^*} & \frac{\partial F_3^*}{\partial x_3^*} \end{vmatrix} \right\}$$

$$- \left(\frac{\partial F_1^*}{\partial x_1^*} + \frac{\partial F_2^*}{\partial x_1^*} + \frac{\partial F_3^*}{\partial x_1^*}\right) \left[x_1^* \left(\frac{\partial F_2^*}{\partial x_2^*} + \frac{\partial F_3^*}{\partial x_3^*}\right) - x_2^* \frac{\partial F_1^*}{\partial x_2^*} - x_3^* \frac{\partial F_1^*}{\partial x_3^*} \right]$$

$$- \left(\frac{\partial F_1^*}{\partial x_2^*} + \frac{\partial F_2^*}{\partial x_2^*} + \frac{\partial F_3^*}{\partial x_2^*}\right) \left[-x_1^* \frac{\partial F_2^*}{\partial x_1^*} + x_2^* \left(\frac{\partial F_1^*}{\partial x_1^*} + \frac{\partial F_3^*}{\partial x_3^*}\right) - x_3^* \frac{\partial F_2^*}{\partial x_3^*} \right]$$

$$- \left(\frac{\partial F_1^*}{\partial x_3^*} + \frac{\partial F_2^*}{\partial x_3^*} + \frac{\partial F_3^*}{\partial x_3^*}\right) \left[-x_1^* \frac{\partial F_3^*}{\partial x_1^*} - x_2^* \frac{\partial F_3^*}{\partial x_2^*} + x_3^* \left(\frac{\partial F_1^*}{\partial x_1^*} + \frac{\partial F_2^*}{\partial x_2^*}\right) \right].$$

Appendix II

The Determinant of the Log-Linear Model

The intermediate steps in deriving the determinant Δ, as given by condition (III.B.2.a.i.2), are as follows:

$$\Delta = \left(\frac{x_1^*}{F_1^*}\right)^2 \left|\begin{matrix} \frac{\alpha_{11}}{x_1^*}F_1^* & \frac{\alpha_{12}}{x_2^*}F_1^* \\ \frac{\alpha_{21}}{x_1^*}F_2^* & \frac{\alpha_{22}}{x_2^*}F_2^* \end{matrix}\right| + \left|\begin{matrix} \frac{\alpha_{11}}{x_1^*}F_1^* & \frac{\alpha_{13}}{x_3^*}F_1^* \\ \frac{\alpha_{31}}{x_1^*}F_3^* & \frac{\alpha_{33}}{x_3^*}F_3^* \end{matrix}\right| + \left|\begin{matrix} \frac{\alpha_{22}}{x_2^*}F_2^* & \frac{\alpha_{23}}{x_3^*}F_2^* \\ \frac{\alpha_{32}}{x_2^*}F_3^* & \frac{\alpha_{33}}{x_3^*}F_3^* \end{matrix}\right|$$

$$-\left(\frac{\alpha_{11}}{x_1^*}F_1^* + \frac{\alpha_{21}}{x_1^*}F_2^* + \frac{\alpha_{31}}{x_1^*}F_3^*\right)\left[x_1^*\left(\frac{\alpha_{22}}{x_2^*}F_2^* + \frac{\alpha_{33}}{x_3^*}F_3^*\right) - x_2^*\frac{\alpha_{12}}{x_2^*}F_1^* - x_3^*\frac{\alpha_{13}}{x_3^*}F_1^*\right]$$

$$-\left(\frac{\alpha_{12}}{x_2^*}F_1^* + \frac{\alpha_{22}}{x_2^*}F_2^* + \frac{\alpha_{32}}{x_2^*}F_3^*\right)$$

$$\times\left[-x_1^*\frac{\alpha_{21}}{x_1^*}F_2^* + x_2^*\left(\frac{\alpha_{11}}{x_1^*}F_1^* + \frac{\alpha_{33}}{x_3^*}F_3^*\right) - x_3^*\frac{\alpha_{23}}{x_3^*}F_2^*\right]$$

$$-\left(\frac{\alpha_{13}}{x_3^*}F_1^* + \frac{\alpha_{23}}{x_3^*}F_2^* + \frac{\alpha_{33}}{x_3^*}F_3^*\right)\left[-x_1^*\frac{\alpha_{31}}{x_1^*}F_3^* - x_2^*\frac{\alpha_{32}}{x_2^*}F_3^*\right.$$

$$\left. + x_3^*\left(\frac{\alpha_{11}}{x_1^*}F_1^* + \frac{\alpha_{22}}{x_2^*}F_2^*\right)\right]$$

$$= \left(\frac{x_1^*}{F_1^*}\right)^2 \left\{\frac{F_1^* \quad F_2^*}{x_1^* \quad x_2^*}\left|\begin{matrix}\alpha_{11} & \alpha_{12} \\ \alpha_{21} & \alpha_{22}\end{matrix}\right| + \frac{F_1^* \quad F_3^*}{x_1^* \quad x_3^*}\left|\begin{matrix}\alpha_{11} & \alpha_{13} \\ \alpha_{31} & \alpha_{33}\end{matrix}\right| + \frac{F_2^* \quad F_3^*}{x_2^* \quad x_3^*}\left|\begin{matrix}\alpha_{22} & \alpha_{23} \\ \alpha_{32} & \alpha_{33}\end{matrix}\right|\right.$$

$$-\frac{1}{x_1^*}(\alpha_{11}F_1^* + \alpha_{21}F_2^* + \alpha_{31}F_3^*)\left[x_1^*\left(\frac{\alpha_{22}}{x_2^*}F_2^* + \frac{\alpha_{33}}{x_3^*}F_3^*\right) - \alpha_{12}F_1^* - \alpha_{13}F_1^*\right]$$

$$-\frac{1}{x_2^*}(\alpha_{12}F_1^* + \alpha_{22}F_2^* + \alpha_{32}F_3^*)\left[-\alpha_{21}F_2^* + x_2^*\left(\frac{\alpha_{11}}{x_1^*}F_1^* + \frac{\alpha_{33}}{x_3^*}F_3^*\right) - \alpha_{23}F_2^*\right]$$

$$\left.-\frac{1}{x_3^*}(\alpha_{13}F_1^* + \alpha_{23}F_2^* + \alpha_{33}F_3^*)\left[-\alpha_{31}F_3^* - \alpha_{32}F_3^* + x_3^*\left(\frac{\alpha_{11}}{x_1^*}F_1^* + \frac{\alpha_{22}}{x_2^*}F_2^*\right)\right]\right\}$$

$$
\begin{aligned}
&= \frac{x_1^*}{x_2^*}\frac{F_2^*}{F_1^*}\begin{vmatrix}\alpha_{11} & \alpha_{12}\\ \alpha_{21} & \alpha_{22}\end{vmatrix} + \frac{x_1^*}{x_3^*}\frac{F_3^*}{F_1^*}\begin{vmatrix}\alpha_{11} & \alpha_{13}\\ \alpha_{31} & \alpha_{33}\end{vmatrix} + \frac{x_1^{*2}}{x_2^* x_3^*}\cdot\frac{F_2^* F_3^*}{F_1^*}\begin{vmatrix}\alpha_{22} & \alpha_{23}\\ \alpha_{32} & \alpha_{33}\end{vmatrix}\\
&\quad - x_1^*\left(\alpha_{11} + \alpha_{21}\frac{F_2^*}{F_1^*} + \alpha_{31}\frac{F_3^*}{F_1^*}\right)\left(\frac{x_1^*}{x_2^*}\frac{F_2^*}{F_1^*}\alpha_{22} + \frac{x_1^*}{x_3^*}\frac{F_3^*}{F_1^*}\alpha_{33} - \alpha_{12} - \alpha_{13}\right)\\
&\quad - \frac{x_1^{*2}}{x_2^*}\left(\alpha_{12} + \alpha_{22}\frac{F_2^*}{F_1^*} + \alpha_{32}\frac{F_3^*}{F_1^*}\right)\left(-\alpha_{21}\frac{F_2^*}{F_1^*} + \frac{x_2^*}{x_1^*}\alpha_{11} + \frac{x_2^*}{x_3^*}\frac{F_3^*}{F_1^*}\alpha_{33} - \alpha_{23}\frac{F_2^*}{F_1^*}\right)\\
&\quad - \frac{x_1^{*2}}{x_3^*}\left(\alpha_{13} + \alpha_{23}\frac{F_2^*}{F_1^*} + \alpha_{33}\frac{F_3^*}{F_1^*}\right)\left(-\alpha_{31}\frac{F_3^*}{F_1^*} - \alpha_{32}\frac{F_3^*}{F_1^*} + \frac{x_3^*}{x_1^*}\alpha_{11} + \frac{x_3^*}{x_2^*}\frac{F_2^*}{F_1^*}\alpha_{22}\right)\\
&= \begin{vmatrix}\alpha_{11} & \alpha_{12}\\ \alpha_{21} & \alpha_{22}\end{vmatrix} + \begin{vmatrix}\alpha_{11} & \alpha_{13}\\ \alpha_{31} & \alpha_{33}\end{vmatrix} + \begin{vmatrix}\alpha_{22} & \alpha_{23}\\ \alpha_{32} & \alpha_{33}\end{vmatrix}\\
&\quad - (\alpha_{11}x_1^* + \alpha_{21}x_2^* + \alpha_{31}x_3^*)(\alpha_{22} + \alpha_{33} - \alpha_{12} - \alpha_{13})\\
&\quad - (\alpha_{12}x_1^* + \alpha_{22}x_2^* + \alpha_{32}x_3^*)(-\alpha_{21} + \alpha_{11} + \alpha_{33} - \alpha_{23})\\
&\quad - (\alpha_{13}x_1^* + \alpha_{23}x_2^* + \alpha_{33}x_3^*)(-\alpha_{31} - \alpha_{32} + \alpha_{11} + \alpha_{22})\\
&= \begin{vmatrix}\alpha_{11} & \alpha_{12}\\ \alpha_{21} & \alpha_{22}\end{vmatrix} + \begin{vmatrix}\alpha_{11} & \alpha_{13}\\ \alpha_{31} & \alpha_{33}\end{vmatrix} + \begin{vmatrix}\alpha_{22} & \alpha_{23}\\ \alpha_{32} & \alpha_{33}\end{vmatrix}\\
&\quad - x_1^*\left\{\begin{vmatrix}\alpha_{11} & \alpha_{12}\\ \alpha_{21} & \alpha_{22}\end{vmatrix} + \begin{vmatrix}\alpha_{11} & \alpha_{13}\\ \alpha_{31} & \alpha_{33}\end{vmatrix} + \begin{vmatrix}\alpha_{12} & \alpha_{13}\\ \alpha_{32} & \alpha_{33}\end{vmatrix} - \begin{vmatrix}\alpha_{12} & \alpha_{13}\\ \alpha_{22} & \alpha_{23}\end{vmatrix}\right\}\\
&\quad - x_2^*\left\{\begin{vmatrix}\alpha_{11} & \alpha_{12}\\ \alpha_{21} & \alpha_{22}\end{vmatrix} + \begin{vmatrix}\alpha_{22} & \alpha_{23}\\ \alpha_{32} & \alpha_{33}\end{vmatrix} + \begin{vmatrix}\alpha_{21} & \alpha_{23}\\ \alpha_{31} & \alpha_{33}\end{vmatrix} + \begin{vmatrix}\alpha_{11} & \alpha_{13}\\ \alpha_{21} & \alpha_{23}\end{vmatrix}\right\}\\
&\quad - x_3^*\left\{\begin{vmatrix}\alpha_{11} & \alpha_{13}\\ \alpha_{31} & \alpha_{33}\end{vmatrix} + \begin{vmatrix}\alpha_{22} & \alpha_{23}\\ \alpha_{32} & \alpha_{33}\end{vmatrix} + \begin{vmatrix}\alpha_{11} & \alpha_{12}\\ \alpha_{31} & \alpha_{32}\end{vmatrix} - \begin{vmatrix}\alpha_{21} & \alpha_{22}\\ \alpha_{31} & \alpha_{32}\end{vmatrix}\right\}\\
&= x_1^*\left\{\begin{vmatrix}\alpha_{22} & \alpha_{23}\\ \alpha_{32} & \alpha_{33}\end{vmatrix} - \begin{vmatrix}\alpha_{12} & \alpha_{13}\\ \alpha_{32} & \alpha_{33}\end{vmatrix} + \begin{vmatrix}\alpha_{12} & \alpha_{13}\\ \alpha_{22} & \alpha_{23}\end{vmatrix}\right\}\\
&\quad - x_2^*\left\{\begin{vmatrix}\alpha_{21} & \alpha_{23}\\ \alpha_{31} & \alpha_{33}\end{vmatrix} - \begin{vmatrix}\alpha_{11} & \alpha_{13}\\ \alpha_{31} & \alpha_{33}\end{vmatrix} + \begin{vmatrix}\alpha_{11} & \alpha_{13}\\ \alpha_{21} & \alpha_{23}\end{vmatrix}\right\}\\
&\quad + x_3^*\left\{\begin{vmatrix}\alpha_{21} & \alpha_{22}\\ \alpha_{31} & \alpha_{32}\end{vmatrix} - \begin{vmatrix}\alpha_{11} & \alpha_{12}\\ \alpha_{31} & \alpha_{32}\end{vmatrix} + \begin{vmatrix}\alpha_{11} & \alpha_{12}\\ \alpha_{21} & \alpha_{22}\end{vmatrix}\right\}
\end{aligned}
$$

References

W. Alonso, 1964, *Location and Land Use*, Harvard University Press.

L. Arnold, R. Lefever (eds.), 1981, *Stochastic Nonlinear Systems in Physics, Chemistry and Biology*, Series in Synergetics, Vol. 8, Springer-Verlag.

W.A. Barnett, E.R. Bernelt, H. White (eds.), 1988, *Dynamic Econometric Modelling*, Cambridge University Press.

D. Campbell, J. Crutchfield, D. Farmer, E. Jen, 1985, "Experimental Mathematics: The Role of Computation in Nonlinear Science," *Communications of the ACM*, **28**:374–384.

J.L. Casti, 1984, "Simple Models, Catastrophes and Cycles," *Kybernetes*, **13**:213–229.

W. Christaller, 1933 (reprinted 1967), *Central Places in Southern Germany* (translated by C.W. Baskin), Prentice-Hall.

J.H. Curry, J.A. Yorke, 1978, "A Transition from Hopf Bifurcation to Chaos: Computer Experiments with Maps in R^2," in N.G. Markley, J.C. Martin, and W. Perrizo (eds.), *The Structure of Attractors in Dynamical Systems*, Springer-Verlag.

R. Day, 1981, "Emergence of Chaos from Neoclassical Growth," *Geographical Analysis*, **13**:315–327.

R. Day, 1982, "Irregular Growth Cycles," *The American Economic Review*, **72**:406–414.

J. Della Dora, J. Demongeot, B. Lacolle (eds.), 1981, *Numerical Methods in the Study of Critical Phenomena*, Series in Synergetics, Vol. 9, Springer-Verlag.

D.S. Dendrinos, 1980, *Catastrophe Theory in Urban and Transport Analysis*, Report DOT/RSPA/DPB-25/80/2, U.S. Department of Transportation, Washington, D.C.

D.S. Dendrinos, 1984a, "Turbulence and Fundamental Urban/Regional Dynamics," Paper presented at the American Association of Geographers Meeting, April 1984, Washington, D.C. Also, at the meeting on "Dynamics Analysis of Spatial Development" at the International Institute for Applied Systems Analysis, Laxenburg, Austria, October 1984.

D.S. Dendrinos, 1984b, "Regions, Antiregions and their Dynamic Stability: The U.S. Case (1929–1979), *Journal of Regional Science*, **24**:65–83.

D.S. Dendrinos, 1984c, "The Structural Stability of the U.S. Regions: Evidence and Theoretical Underpinnings," *Environment and Planning A*, **16**:1433–1443.

D.S. Dendrinos, H. Mullally, 1981, "Evolutionary Patterns of Urban Populations," *Geographical Analysis*, **13**:328–344.

D.S. Dendrinos (with H. Mullally), 1985, *Urban Evolution: Studies in the Mathematical Ecology of Cities*, Oxford University Press.

D.S. Dendrinos, M. Sonis, 1986, "Variational Principles and Conservation Conditions in Volterra's Ecology and in Urban Relative Dynamics," *Journal of Regional Science*, **26**:359–377.

D.S. Dendrinos, M. Sonis, 1987, "The Onset of Turbulence in Discrete Relative Multiple Spatial Dynamics," *Journal of Applied Mathematics and Computation*, **22**:25–44.

D.S. Dendrinos, M. Sonis, 1988, "Nonlinear Relative Discrete Population Dynamics of the U.S. Regions," *Journal of Applied Mathematics and Computation*, **23**:265–285.

M. Eigen, P. Schuster, 1979, *The Hypercycle: A Principle of Natural Self-Organization*, Springer-Verlag.

M. Feigenbaum, 1978, "Quantitative Universality for a Class of Nonlinear Transformations," *Journal of Statistical Phyics*, **19**:25–52.

M. Feigenbaum, 1979, "The Onset Spectrum of Turbulence," *Physics Letters*, **74A**:375.

M. Feigenbaum, 1980, "Universal Behavior in Nonlinear Systems," *Los Alamos Science*, Summer:4–27.

E. Frehland (ed.), 1984, *Synergetics: from Microscopic to Macroscopic Order*, Series in Synergetics, Vol. 22, Springer-Verlag.

L. Garrido, 1983, *Dynamical Systems and Chaos*, Lecture Notes in Physics, Vol. 179, Springer-Verlag.

M.E. Gilpin, 1979, "Spiral Chaos in a Predator–Prey Model," *American Naturalist*, **113**:306–8.

J. Guckenheimer, P. Holmes, 1983, *Nonlinear Oscillations, Dynamical Systems and Bifurcations in Vector Fields*, Springer-Verlag.

O. Gurel, O.E. Rossler (eds.), 1979, *Bifurcation Theory and Applications in Scientific Disciplines*, Annals of the New York Academy of Sciences, Vol. 316.

H. Haken, 1977, *Synergetics; an Introduction*, Series in Synergetics, Vol. 1, Springer-Verlag.

H. Haken (ed.), 1981, *Chaos and Order in Nature*, Series in Synergetics, Vol. 11, Springer-Verlag.

H. Haken, 1983, *Advanced Synergetics*, Series in Synergetics, Vol. 20, Springer-Verlag.

B. Hannon, 1985, "Linear Dynamic Ecosystems," *Journal of Theoretical Biology*, **116**:89–110.

B. Harris, A.G. Wilson, 1978, "Equilibrium Values and Dynamics of Attractiveness Terms in Production-Constrained Spatial-Interaction Models," *Environment and Planning A*, **10**:371–388.

R.H.G. Helleman, 1981, "Feigenbaum Sequences in Conservative and Dissipative Systems," in H. Haken (ed.), *Chaos and Order in Nature*, pp. 232–248, Springer-Verlag.

A.V. Holden (ed.), 1986, *Chaos*, Princeton University Press.

C.W. Horton, L.E. Reichl, V.G. Szebehely (eds.), 1983, *Long-Time Prediction in Dynamics*, Wiley Series on Non-Equilibrium Problems in the Physical Sciences and Biology (eds. I. Prigogine and G. Nicolis).

H. Hotelling, 1929, "Stability in Competition," *Economic Journal*, **XXXIX**:41–57.

G. Iooss, R.H.G. Helleman, R. Stora, 1983, *Chaotic Behavior of Deterministic Systems*, North-Holland.

W. Isard, 1956, *Location and Space Economy*, MIT Press.

W. Isard, 1975, *Introduction to Regional Science*, Prentice-Hall.

V. Leontief, 1970, in A. Carter and A. Brady (eds.), *Contributions to Input–Output Analysis*, Vol. 1, Ch. 1, North-Holland.

T.Y. Li, J.A. Yorke, 1975, "Period Three Implies Chaos," *American Mathematical Monthly*, **82**:985–992.

E.N. Lorenz, 1963, "Deterministic Non-periodic Flow," *Journal of Atmospheric Sciences*, **20**:130–141.

A. Lösch, 1937, *The Economics of Location* (translated by W. Woglum and V.A. Stolpes), Yale University Press, 1954.

R.E. Lucas, 1981, *Studies in Business-Cycle Theory*, MIT Press.

J.E. Marsden, M. McCraken, 1976, *The Hopf Bifurcation and its Applications*, Applied Mathematics Sciences, Vol. 19, Springer-Verlag.

R. May, 1974, "Biological Populations with Nonoverlapping Generations: Stable Points, Stable Cycles, and Chaos," *Science*, **186**:645–647.

R. May, 1976, "Simple Mathematical Models with Very Complicated Dynamics," *Nature*, **261**:459–467.

R. May, A. Oster, 1976, "Bifurcation adn Dynamic Complexity in Simple Ecological Models," *The American Naturalist*, **110**:573–599.

G. Nicolis, I. Prigogine, 1977, *Self-Organization in Non-Equilibrium Systems: from Dissipative Structures to Order Through Fluctuations*, Wiley-Interscience.

G. Nicolis, G. Dewel, J.W. Turner (eds.), 1981, *Order and Fluctuations in Equilibrium and Nonequilibrium Statistical Mechanics*, XVI International Solvay Conference on Physics. Wiley Series on Non-Equilibrium Problems in the Physical Sciences and Biology (eds. I. Prigogine and G. Nicolis).

P. Nijkamp, 1985, "Long-Term Economic Fluctuations: A Spatial View," mimeo, Department of Economics, Free University of Amersterdam.

P. Nijkamp (ed.), 1986, *Technological Change, Employment and Spatial Dynamics*, Springer-Verlag.

T. Poston, I. Stewart, 1978, *Catastrophe Theory and its Applications*, Fearon-Pitman.

T. Poston, A.G. Wilson, 1977, "Facility Size vs. Distance Travelled: Urban Services and the Fold Catastrophe," *Environment and Planning A*, **9**:681–686.

R. Reiner, M. Munz, G. Haag, W. Weidlich, 1986, "Chaotic Evolution of Migratory Systems," *Sistemi Urbani*, **2/3**:285–308.

T.L. Saaty, 1981, *Modern Nonlinear Equations*, Dover.

P.A. Samuelson, 1941, "Conditions that the Roots of a Polynomial Be Less than Unit in Absolute Value," *Annals of Mathematical Statistics*: 360–364.

W.M. Schaffer, G.L. Truty, S.L. Fulmer, 1988, *Dynamical Software*. Dynamical Systems, Inc.

P. Schuster (ed.), 1984, *Stochastic Phenomena and Chaotic Behavior in Complex Systems*, Series in Synergetics, Vol. 21, Springer-Verlag.

A.C. Scott, F.Y.F. Chu, D.W. McLaughin, 1973, "The Soliton: A New Concept in Applied Science," *Proc. IEEE*, **61**:1443–1483.

F.M. Scudo, J.R. Ziegler, 1978, *The Golden Age of Theoretical Ecology: 1923–1940*, Lecture Notes in Biomathematics Series, Vol. 22, Springer-Verlag.

S. Smale, 1973, "Stability and Isotropy in Discrete Dynamical Systems," in M. Peixoto (ed.), *Dynamical Systems*, Academic Press.

M. Sonis, 1982, "The Decomposition Principle vs. Optimization in Regional Analysis: The Inverted Problem of Multiobjective Programming, in G. Chiotis, D. Tsoukalas, and H. Louri (eds.), *The Regions and the Enlargement of the European Economic Community*, pp. 35–60, Athens: Eptalofos.

M. Sonis, 1983a, "Competition and Environment: A Theory of Temporal Innovation Diffusion" in D.A. Griffith and A.C. Lea (eds.), *Evolving Geographical Structures*, pp. 99–129, Martinus Nijhoff.

M. Sonis, 1983b, "Spatio-Temporal Spread of Competitive Innovations: an Ecological Approach," *Papers of the Regional Science Association*, **52**: 159–174.

M. Sonis, 1984, "Dynamic Choice of Alternatives, Innovation Diffusion and Ecological Dynamics of Volterra–Lotka Model," *London Papers in Regional Science*, **14**: 29–43.

M. Sonis, 1985, "Unifying Principles in Spatial Analysis—A Search for Theory," Paper presented at the Conference on Scientific Geography, University of Georgia, Athens, Georgia, March 1985.

M. Sonis, 1986a, *Quantitative and Qualitative Methods for Relative Spatial Dynamics*, CERUM Lectures in Regional Science, Nordic Workshop in Regional Science, April 15–20, 1985, University of Umea, Sweden.

M. Sonis, 1986b, "A Unified Theory of Innovation Diffusion, Dynamic Choice of Alternatives, Ecological Dynamics and Urban/Regional Growth and Decline," *Ricerche Economiche*, **XL**: 696–723.

M. Sonis, 1987, "Regional Growth, Regional Decline and Decentralization," in P. Freidrich and I. Masser (eds.), *International Perspectives on Regional Decentralization*, Nomos, Baden-Baden.

M. Sonis, 1988, "Relationships between Economic and Spatial Analysis: Methodological Discussion," *Horizons*, The Haifa University, Vol. 23–24: 122–115.

M. Sonis, 1989, "Discrete Time Logistic Growth," mimeo, Bar-Ilan University.

M. Sonis, D.S. Dendrinos, 1987a, "A Discrete Relative Growth Model: Switching, Role Reversal and Turbulence," in P. Friedrich and I. Masser (eds.), *International Perspectives of Regional Decentralization*, "Schriften zur offentlichen Verwaltung und offentlichen Wirtschaft," Vol. 87, Nomos, Baden-Baden.

M. Sonis, D.S. Dendrinos, 1987b, "Period-Doubling in Discrete Relative Spatial Dynamics and the Feigenbaum Sequence," *Mathematical Modeling: An International Journal*, **9**: 539–546.

M. Sonis, D.S. Dendrinos, 1988, "Multiple-Stock, Multiple-Location Volterra–Lotka Dynamics are Degenerate," *Sistemi Urbani* (forthcoming 1990).
sociation Meeting, Cambridge, August 1989.

C. Sparrow, 1982, *The Lorenz Equations: Bifurcations, Chaos and Strange Attractors*, Applied Mathematical Sciences, Vol. 41, Springer-Verlag.

R. Thom, 1975, *Structural Stability and Morphogenesis*, W.A. Benjamin.

J.H. von Thunen, 1826, *Der isolierte Staat in Beziehung auf Nationalekonomie und Landwistschaft*, reprinted by A. Fischer in 1966.

V. Volterra, 1927, "Variations and Fluctuations in the Number of Coexisting Animal Species," in F.M. Scudo and J.R. Ziegler (eds.), *The Golden Age of Theoretical Ecology: 1923–1940*, Lecture Notes in Biomathematics, Vol. 22, Springer-Verlag.

W. Weidlich, G. Haag, 1983, *Quantitative Sociology; The Dynamics of Interacting Populations*, Synergetics Series, Vol. 19, Springer-Verlag.

W. Weidlich, G. Haag (eds.), 1988, *Interregional Migration—Dynamic Theory and Comparative Analysis*, Springer-Verlag.

A.G. Wilson, 1981 *Catastrophe Theory and Bifurcation: Applications to Urban and Regional Systems*, Croom-Helm.

K. G. Wilson, 1971, *Physical Review*, **B4**: 3174–3184.

E.C. Zeeman, 1977, *Catastrophe Theory; Selected Papers, 1972–77*, Addison-Wesley.

Author Index

Subject Index

Applied Mathematical Sciences

52. *Chipot:* Variational Inequalities and Flow in Porous Media.
53. *Majda:* Compressible Fluid Flow and Systems of Conservation Laws in Several Space Variables.
54. *Wasow:* Linear Turning Point Theory.
55. *Yosida:* Operational Calculus: A Theory of Hyperfunctions.
56. *Chang/Howes:* Nonlinear Singular Perturbation Phenomena: Theory and Applications.
57. *Reinhardt:* Analysis of Approximation Methods for Differential and Integral Equations.
58. *Dwoyer/Hussaini/Voigt (eds.):* Theoretical Approaches to Turbulence.
59. *Sanders/Verhulst:* Averaging Methods in Nonlinear Dynamical Systems.
60. *Ghil/Childress:* Topics in Geophysical Dynamics: Atmospheric Dynamics, Dynamo Theory and Climate Dynamics.
61. *Sattinger/Weaver:* Lie Groups and Algebras with Applications to Physics, Geometry, and Mechanics.
62. *LaSalle:* The Stability and Control of Discrete Processes.
63. *Grasman:* Asymptotic Methods of Relaxation Oscillations and Applications.
64. *Hsu:* Cell-to-Cell Mapping: A Method of Global Analysis for Nonlinear Systems.
65. *Rand/Armbruster:* Perturbation Methods, Bifurcation Theory and Computer Algebra.
66. *Hlaváček/Haslinger/Nečas/Lovíšek:* Solution of Variational Inequalities in Mechanics.
67. *Cercignani:* The Boltzmann Equation and Its Applications.
68. *Temam:* Infinite Dimensional Dynamical System in Mechanics and Physics.
69. *Golubitsky/Stewart/Schaeffer:* Singularities and Groups in Bifurcation Theory, Vol. II.
70. *Constantin/Foias/Nicolaenko/Temam:* Integral Manifolds and Inertial Manifolds for Dissipative Partial Differential Equations.
71. *Catlin:* Estimation, Control, and the Discrete Kalman Filter.
72. *Lochak/Meunier:* Multiphase Averaging for Classical Systems.
73. *Wiggins:* Global Bifurcations and Chaos.
74. *Mawhin/Willem:* Critical Point Theory and Hamiltonian Systems.
75. *Abraham/Marsden/Ratiu:* Manifolds, Tensor Analysis, and Applications, 2nd ed.
76. *Lagerstrom:* Matched Asymptotic Expansions: Ideas and Techniques.
77. *Aldous:* Probability Approximations via the Poisson Clumping Heuristic.
78. *Dacorogna:* Direct Methods in the Calculus of Variations.
79. *Hernández-Lerma:* Adaptive Markov Control Processes.
80. *Lawden:* Elliptic Functions and Applications.
81. *Bluman/Kumei:* Symmetries and Differential Equations.
82. *Kress:* Linear Integral Equations.
83. *Bebernes/Eberly:* Mathematical Problems from Combustion Theory.
84. *Joseph:* Fluid Dynamics of Viscoelastic Liquids.
85. *Yang:* Wave Packets and Their Bifurcations in Geophysical Fluid Dynamics.
86. *Dendrinos/Sonis:* Chaos and Socio-Spatial Dynamics.